Beginning AutoCAD 2010

EXERCISE WORKBOOK

by
Cheryl R. Shrock

Professor
Drafting Technology
Orange Coast College, Costa Mesa, Ca.
Autodesk Authorized Author

INDUSTRIAL PRESS
New York

Autodesk, AutoCAD, Design Web Format, and DWF are either registered
trademarks or trademarks of Autodesk, Inc., in the U.S.A. and / or certain other
countries. Certain content, including trial software, provided courtesy of Autodesk,
Inc., © 2009. All rights reserved.

Industrial Press Inc.
989 Avenue of the Americas
New York, NY 10018

10 9 8 7 6 5 4 3 2

This book is dedicated to the teachers that continue to use my workbooks within their AutoCAD courses and the independent users that seek an easy to use instruction book to guide them through their introduction to AutoCAD.

You are the reason I write these workbooks.

AutoCAD Books by Cheryl R. Shrock:

Beginning AutoCAD **2006**	ISBN 978-08311-3213-2
Advanced AutoCAD **2006**	ISBN 978-08311-3214-9
Beginning AutoCAD **2007**	ISBN 978-08311-3302-3
Advanced AutoCAD **2007**	ISBN 978-08311-3303-0
Beginning AutoCAD **2008**	ISBN 978-0-8311-3341-2
Advanced AutoCAD **2008**	ISBN 978-0-8311-3342-9
Beginning AutoCAD **2009**	ISBN 978-0-8311-3359-7
Advanced AutoCAD **2009**	ISBN 978-0-8311-3360-3
Beginning AutoCAD **2010**	ISBN 978-0-8311-3404-4
Advanced AutoCAD **2010**	ISBN 978-0-8311-3400-6
AutoCAD Pocket Reference **2006,** Releases 2006-2004	ISBN 978-0-8311-3263-7
AutoCAD Pocket Reference **2007** Releases 2007-2004	ISBN 978-0-8311-3328-3
AutoCAD Pocket Reference **2008** Releases 2008-2006	ISBN 978-0-8311-3354-2
AutoCAD Pocket Reference **4th** Edition, Release 2009 or later	ISBN 978-0-8311-3384-9

For information about these books visit www.industrialpress.com

For information about Cheryl Shrock's online courses, visit www.shrockpublishing.com

Table of Contents

INTRODUCTION

About this workbook
Beginning AutoCAD® 2010, Exercise Workbook is designed for classroom instruction or self-study. There are 30 lessons. Each lesson starts with step-by-step instructions followed by exercises designed for practicing the commands you learned within that lesson.

You may find the order of instruction in this workbook somewhat different from most textbooks. The approach I take is to familiarize you with the drawing commands first. After you are comfortable with the drawing commands, you will be taught to create your own setup drawings. This method is accomplished by supplying you with a drawing, "**2010-Workbook Helper.dwg**". This drawing is preset and ready for you to open and use. For the first 8 lessons you should not worry about settings, **you just draw.**

Important
How to get the drawing listed above?
The file "2010-Workbook Helper" should be downloaded from our website:
www.shrockpublishing.com

Printing
The exercises in the workbook have been designed to be printed on any **8-1/2 X 11 (Letter size) printer**. You will be able to print your exercises and view your results as you progress through the lessons. If you would like to configure a large format printer refer to **Appendix-A to "Add a Printer / Plotter"**.

About the Author
Cheryl R. Shrock is a Professor and Chairperson of Computer Aided Design at Orange Coast College in Costa Mesa, California. She is also an Autodesk® registered author. Cheryl began teaching CAD in 1990. Previous to teaching, she owned and operated a commercial product and machine design business where designs were created and documented using CAD. This workbook is a combination of her teaching skills and her industry experience.

The author can be contacted for questions or comments at: *learnacad@yahoo.com*

"Sharing my industry and CAD knowledge has been the most rewarding experience of my career. Students come to learn CAD in order to find employment or to upgrade their skills. Seeing them actually achieve their goals, and knowing I helped, is a real pleasure. If you read the lessons and do the exercises, I promise, you will not fail."

Cheryl R. Shrock

CONFIGURING YOUR SYSTEM

AutoCAD ® allows you to customize it's configuration. While you are using this workbook it is necessary for you to make some simple changes to your configuration so our configurations are the same. This will ensure that the commands and exercises work as expected. The following instructions will guide you through those changes.

1. Start AutoCAD®
    a. To start AutoCAD, follow the instructions on page 1-2.
        Then return to this page and go to step 2.

2. Type: **_options_** then press the **<enter>** key. (not case sensitive)

    Depending how your software is set up, the text that you type will either appear on the Command Line, as shown below on the left, or in the Dynamic Input box, as shown below on the right. Either is fine for now and will be discussed later.

Command Line   **Dynamic Input**

NOTE:

AUTOCAD LT USERS:
You may find that some of the settings appear slightly different.
But they are mostly the same.

Configuration Settings

3. Select the **Display** tab and change the settings on your screen to match the dialog box below.

4. Select the **Open and Save** tab and change the settings on your screen to match the dialog box below.

Configuration Settings

5. Select the *Plot and Publish* tab and change the settings on your screen to match the dialog box below.

You may use this setting or select the actual name of the printer

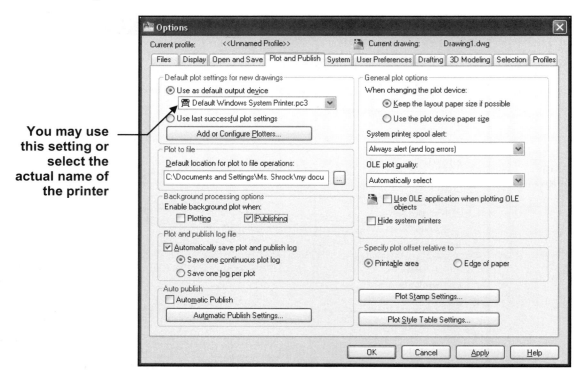

6. Select the *System* tab and change the settings on your screen to match the dialog box below.

Configuration Settings

7. Select the **User Preferences** tab and change the settings on your screen to match the dialog box below.

Rt-click Cust.
Select this button and change the settings shown below

LW settings.
Select this button and change the settings shown below

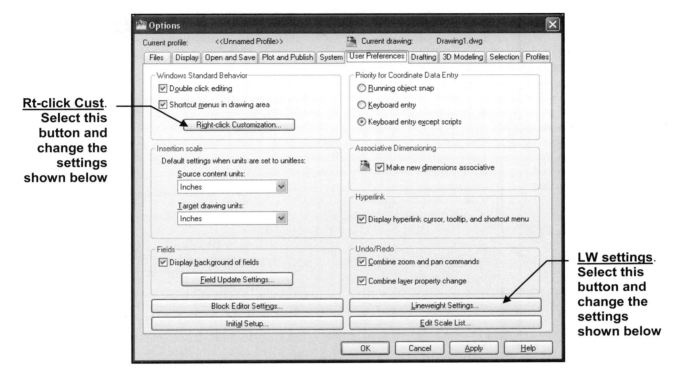

8. After making the setting changes shown below select **Apply & Close** button.

Right-click Customization

Lineweight Settings

Configuration Settings

9. Select the **Drafting** tab and change the settings on your screen to match the dialog box below.

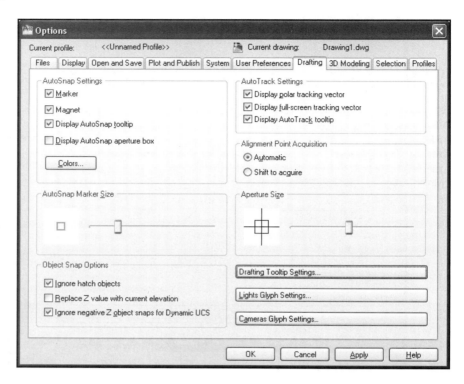

10. Select the **Selection** tab and change the settings on your screen to match the dialog box below. (Note: 3D Modeling tab was skipped)

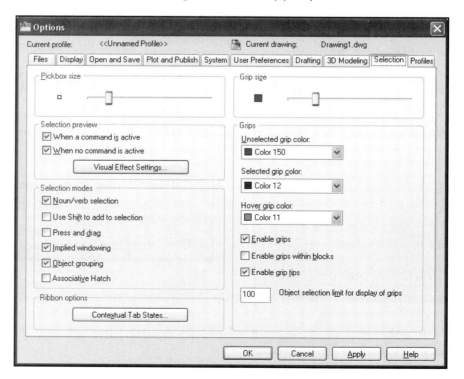

Configuration Settings

11. Select the **Apply** button.

12. Select the **OK** button.

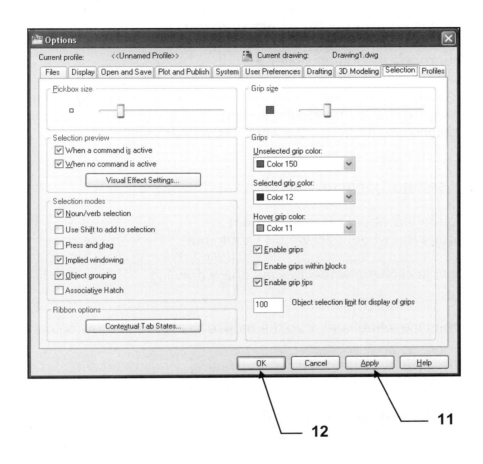

13. Now you should be back to the AutoCAD screen. But there is one more step.

It is important that you configure your AutoCAD software to only allow one drawing open on the screen at one time. It will be less confusing for now. When you progress to the Advanced Workbook you will configure AutoCAD to allow multiple drawings on the screen. But for now please do the following:

1. Type **SDI** and press **<enter>**
2. Type **1** and press **<enter>**

*Note: If the command line says **"Invalid"**, you have multiple files already open. Close all but one and try again.*

Customizing your Wheel Mouse

A Wheel mouse has two or more buttons and a small wheel between the two topside buttons. The default functions for the two top buttons and the Wheel are as follows:
Left Hand button is for **input** and can't be reprogrammed.
Right Hand button is for **Enter** or the **shortcut menu**.
The Wheel may be used to <u>Zoom and Pan</u> or <u>Zoom and display</u> the Object Snap menu. You will learn more about this later.

The following describes how to select the Wheel functions. After you understand the functions, you may choose to change the setting.
To change the setting you must use the **MBUTTONPAN** variable.

MBUTTONPAN setting 1: (Factory setting)

ZOOM   Rotate the wheel forward to zoom in
                     Rotate the wheel backward to zoom out

ZOOM   Double click the wheel to view entire drawing
EXTENTS

PAN   Press the wheel and drag the mouse to move the drawing on the screen.

MBUTTONPAN setting 0:

ZOOM   Rotate the wheel forward to zoom in
                     Rotate the wheel backward to zoom out

OBJECT   Object Snap menu will appear when you press the wheel
SNAP

To change the setting:

1. Type: **mbuttonpan <enter>**
2. Enter **0** or **1 <enter>**

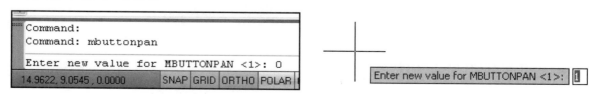

Command Line   **Dynamic Input**

AutoCAD 2010 System Requirements

AutoCAD 2D and AutoCAD LT – <u>32-bit</u>

Operating system:
Microsoft Windows Vista, any of the versions
Microsoft Windows XP Home or Professional SP 2

Browser
Microsoft Internet Explorer 6.0 or later

Memory and Hard Disk Space
XP-1 GB of RAM minimum
Vista - 2GB of Ram minimum
750MB free hard disk space for installation

Hardware (required)
XP - AMD 64-based or Intel Pentium 4, 1.6 Ghz or higher
Vista - AMD 64-based or Intel Pentium 4, 3.0 Ghz or higher
Pointing Device - MS-Mouse compatible
Display - 1024 X 768 VGA (True color)
DVD/CD-ROM drive for initial installation only

> **Note:**
> Whether the Windows operating system is the 32-bit or the 64-bit version is automatically detected when installing AutoCAD. The appropriate version of AutoCAD will be installed. The 32-bit version of AutoCAD cannot be installed on a 64-bit version of Windows.

AutoCAD <u>2D</u> – <u>64-bit additional</u> requirements

Operating system:
Microsoft Windows Vista, 64-bit
Microsoft Windows XP Professional x64

Memory and Hard Disk Space
XP Pro - 1GB RAM
Vista 64-bit – 2GB RAM

Hardware (required)
Intel 64-based or AMD 64 Processor

AutoCAD <u>3D</u> – 32 or 64-bit <u>additional minimum</u> requirements

Operating system:
Microsoft Windows Vista
Microsoft Windows XP Professional

Memory and Hard disk space
2GB RAM
2GB free hard disk
1GB free hard disk for installation (in addition to the 2GB free hard disk above)

Hardware (required)
Intel Pentium 4.0 GHz Processor or AMD Athion processor, 2.2 Ghz or greater or
Intel or AMD Dual Core processor, 1.6 Ghz or greater.
XP Pro - 1280 X 1024 32-bit color video display adapter, True Color, 128 MB, or Direct3D
capable workstation class graphics card.
Vista - Direct3D capable workstation class graphics card with 128MB
1024 X 768 VGA with True Color (minimum)

Go to Autodesk.com to verify graphics hardware for use with AutoCAD 2010

Notes:

LEARNING OBJECTIVES

After completing this lesson, you will be able to:

1. Start AutoCAD program
2. Understand AutoCAD Workspaces
3. Recognize all of the features in the AutoCAD Window
4. Use AutoCAD Help

LESSON 1

STARTING AUTOCAD

To Start AutoCAD use one of the 2 methods below.
(Be patient it may take a few minutes to load. It is a large program)

1. Select **Start / All Programs / Autodesk / AutoCAD 2010**
    or
2. Double click on the AutoCAD 2010 desktop shortcut icon.

    Initial Setup should appear
    Initial setup allows you to customize AutoCAD when you start the first time. After responding to a few questions AutoCAD will add a few tools and palettes that are specific to your industry.

    You may experiment with after you are more familiar with AutoCAD.
    <u>For now I would like you to select **Skip**</u>.

    You may return to Initial Setup any time by selecting
    Options / User Preferences tab   Refer to page Intro-5

3. Select the **Skip** button.

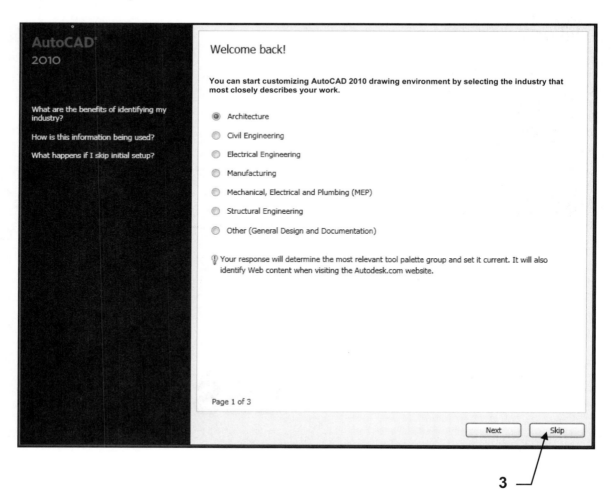

STARTING AUTOCAD....continued

Skip Initial Setup should appear

4. Select the ☑ **Remind me the next time AutoCAD 2010 starts.**

5. Select **Start AutoCAD 2010 button**.

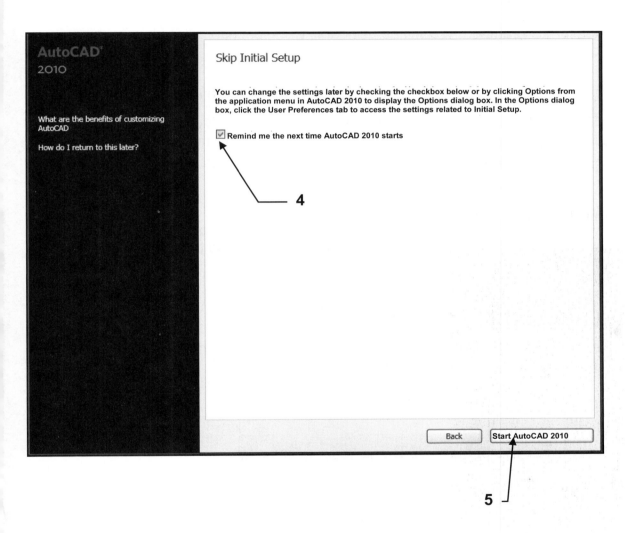

AutoCAD Workspaces

WHAT IS A WORKSPACE?

Workspaces controls the display of menus, toolbars, and palettes.
AutoCAD gives you the option of deciding how you would like them displayed.
There are 3 preset Workspaces.

2D Drafting & Annotation (shown on the next page)
This workspace is the default display. It displays the necessary menus, tools and palettes for 2D drafting. We will be using this workspace for all lessons within this workbook.

AutoCAD Classic or AutoCAD LT Classic
AutoCAD kept the classic workspace from older releases of AutoCAD if a user preferred not to use the Ribbon. The classic workspace arranges the toolbars on the sides and top of the screen.

3D Modeling
This workspace will display the necessary menus, tools and palettes for 3D drafting. This workspace will not be used with this workbook because this workbook does not include 3D commands.

HOW TO SELECT A WORKSPACE

1. Selecting a workspace is easy. Select the ▼ located in the lower right corner of the screen.

2. Select the workspace from the pop up menu that appears.

Note: We will be using:
    2D Drafting & Annotation.

The "check" box ☑ indicates which workspace is current and in use.

AUTOCAD APPLICATION WINDOW

The AutoCAD Application Window default workspace
2D Drafting & Annotation, is shown below.

Important:
If your screen does not appear as shown below go to the Intro section and
follow the steps for configuring AutoCAD to match the workbook configuration.

A description of each area is shown on the following pages.

APPLICATION WINDOW DESCRIPTIONS

APPLICATION MENU

The Application Menu provides easy access to common tools
Each of the tools will be discussed later in the workbook.

1. Click on the **Application Menu button** in the upper left corner of the AutoCAD display screen.

1. Application Menu button

Open Documents and **Recent Documents** are described on the following pages.

These will be discussed later

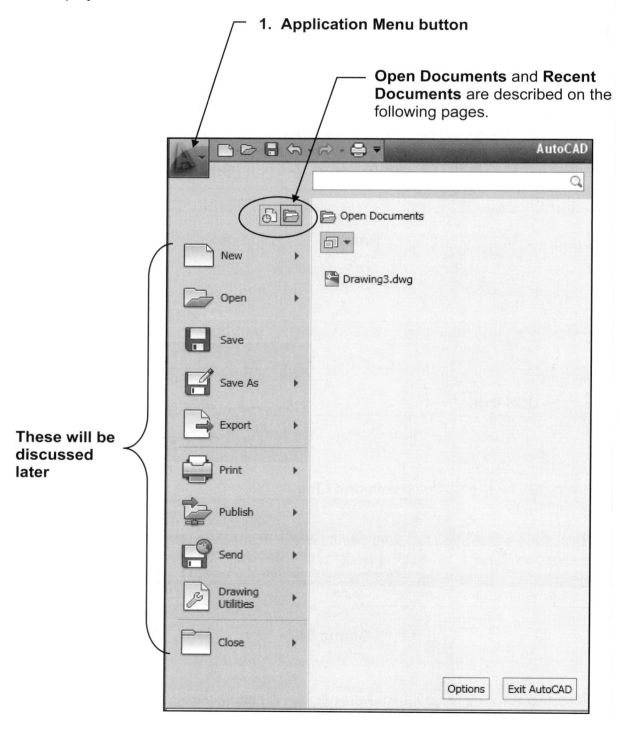

APPLICATION WINDOW DESCRIPTIONS....continued

Open Documents

First let me emphasis **this is not a method to "open" a drawing file**.

Open Documents is a list of all documents that are **already open** within AutoCAD.

This list may be displayed as icons or images
If you hover the cursor over a document name a preview image will appear.

Open documents button

Display choice

List of open documents

Important setting change for new AutoCAD users.

AutoCAD allows you to have multiple drawing files open at the same time.
This is very helpful to experienced AutoCAD users but can be confusing for users NEW to AutoCAD.

While using this workbook you might find it helpful if you change a setting to restrict you to only one drawing file to be open at one time. This will only change AutoCAD and will have no affect on any other software on your computer. In addition this setting may be turned ON and OFF instantly.

1. You should only have one drawing open. Close all but one drawing file.
2. Type: sdi <enter>
3. Type: 1 <enter>

Now AutoCAD will restrict you to one drawing on the screen. If you choose to go back to multiple drawings repeat the steps above except enter "0" instead of "1".

APPLICATION WINDOW DESCRIPTIONS....continued

View Recent Documents

When you select **Recent Documents** a list of the recently viewed documents will appear.

This list may be displayed as icons or images and may be sorted in an ordered list or grouped by date or file type.

If you hover the cursor over a document name a preview image will appear.

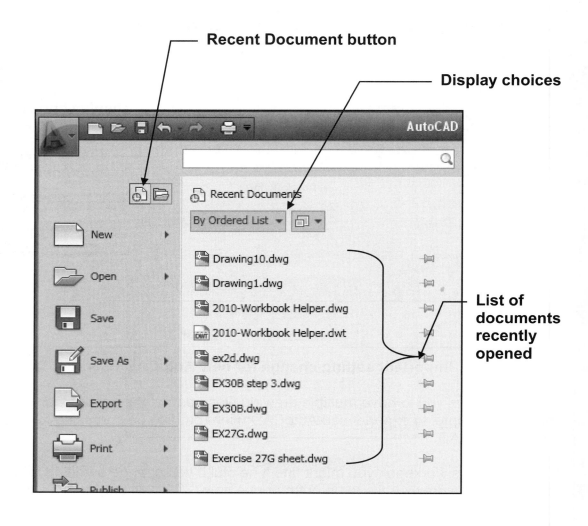

APPLICATION WINDOW DESCRIPTIONS....continued

Quick Access Toolbar

The **Quick Access Toolbar** is located in the top left corner of the AutoCAD window. It includes the most commonly used tools, such as New, Open, Save, Undo and Redo and Plot.

Customize the Quick Access Toolbar
You can add tools with the Customize User Interface dialog box.

For Example:
I like students new to AutoCAD to use the "**Save As**" commandl rather than the "**Save**" command. So I add the "**Save As**" tool to the Quick Access Toolbar. If you would like to add the "**Save As**" tool to your Quick Access Toolbar follow the steps below.

1. Place the Cursor on the Quick Access Toolbar and press the right mouse button.

2. Select "**Customize Quick Access Toolbar...**" from the pop up menu.

3. Scroll through the list of Commands to "**Save As**..."

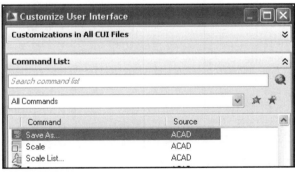

4. Press the Left mouse button on "Save As..." and drag it to the Quick Access Toolbar and drop it by releasing the left mouse button.

5. Select the **OK** button at the bottom of the Customize User Interface dialog box.

The Customize User Interface dialog box will disappear and the new Quick Access Toolbar is saved to the current Workspace.

To Remove a tool:
Place the cursor on the top and press the right mouse button.
Select **Remove from Quick Access Toolbar.**

APPLICATION WINDOW DESCRIPTIONS....continued

RIBBON

The RIBBON provides access to the AutoCAD tools.
The TABS contain multiple PANELS. Each PANEL contains multiple tools.
When you select a TAB a new set of PANELS will appear.

Additional Tools

If you select the ▼ symbol, at the bottom of the panel, the panel will expand to access additional tools.

Control the display of Tabs and Panels
Right click on the Ribbon and select which Tabs or Panels you choose to display.

Control the TAB order
If you would like to change the order of the tabs click and drag the tab to the new location.

Floating PANELS
If you prefer to separate a Panel from the Ribbon you may drag the panel off the Ribbon to a new location on the screen.

APPLICATION WINDOW DESCRIPTIONS....continued

STATUS BAR

The Status Bar is located on the bottom of the screen. It displays the current settings. These settings can be turned ON or OFF by clicking on one of the buttons or by pressing a corresponding function key, F1, F2 etc.

Left Side buttons

Coordinate Display

**Right Side buttons
(Not discussed at
this time)**

Coordinate Display

These numbers will change as you move the cursor. The coordinates display the location of the cursor in reference to the Origin. The Origin is currently in the lower left corner of the Drawing Area.

The first number represents the horizontal movement (X-axis), the second number represents the vertical movement (Y-axis) and the third number is the Z-axis which is used for 3D.

This is basically for reference and will not be used often.

Left side buttons
The left side of the Status bar displays 10 icons.
This display can be changed from **icons** to **traditional text labels**.

ICONS   **TRADITIONAL TEXT LABELS**

Note: I think it will make it easier to learn if you make this change.

To change the icon display to traditional text labels.
Right click on any of the icons and select "**use icons**".

The check mark means use **icons**.
No check mark means use **traditional text labels**.

A brief description of each button is shown on the next page.
Each button will be discussed in more detail in future Lessons.

APPLICATION WINDOW DESCRIPTIONS....continued

STATUS BAR BUTTON DESCRIPTIONS

| SNAP | GRID | ORTHO | POLAR | OSNAP | OTRACK | DUCS | DYN | LWT | QP |

ICONS   **TRADITIONAL TEXT LABELS**

SNAP (You may also use F9 to toggle ON or OFF)
Increment Snap controls the incremental movement of the cursor. If it is ON the cursor will "snap" in an incremental movement. If it is OFF the cursor will move smoothly. You may set the increments by right clicking on the Snap button and selecting Settings.

GRID (You may also use F7 to toggle ON or OFF)
The dots in the Drawing Area are called the Grid. It is only a drawing aid and will not print. The default spacing is 1 unit of measurement. You may change the Grid spacing at any time by right clicking on the Grid button and selecting Settings.

ORTHO (You may also use F8 to toggle ON or OFF)
Ortho restricts the movement of the cursor to Horizontal or Vertical. When Ortho is ON the cursor moves only in the horizontally or vertically. When Ortho is OFF the cursor moves freely in any direction.

POLAR (You may also use F10 to toggle ON of OFF)
PolarSnap restricts cursor movement to specified increments along a polar angle.

OSNAP (You may also use F3 to toggle ON of OFF)
When Osnap is ON the cursor will "snap" to preset locations on objects.

OTRACK (You may also use F11 to toggle ON of OFF)
Polar tracking restricts cursor movement to specified angles.

DUCS (You may also use F6 to toggle ON of OFF)
Dynamic User Coordinate System changes the grid plane to follow the XY plane of the dynamic UCS. Used for 3D, refer to Exercise Workbook for Advanced AutoCAD.

DYN (You may also use F12 to toggle ON of OFF)
When Dynamic Input is on, you can enter coordinate values in tooltips near the cursor.

LWT
Lineweight displays the width assigned to each object. When it is ON the lineweights are visible.

QP
If ON, Quick Properties displays the properties of the object selected. If OFF the Quick Properties box will not appear.

APPLICATION WINDOW DESCRIPTIONS....continued

COMMAND LINE

The **Command Line** is located at the bottom of the screen. This is where AutoCAD will prompt you for information and you will enter commands, values and select options. Basically this is how you communicate with AutoCAD.

When you select a tool such as Circle ⊙ ▾ the prompt and options will be displayed on the command line.

Command   **Prompt**

```
Command: *Cancel*
Command: c

CIRCLE Specify center point for circle or [3P/2P/Ttr (tan tan radius)]:
```

AutoCAD then responds with a "*prompt*" that asks you to:

"*Specify center point for circle*" or *[3P/2P/Ttr (tan tan radius)]*:

The information within the [ ] brackets are options that you may select.
This will be discussed more in Lesson 2.

Some users prefer to type their command choice on the command line rather than selecting a tool or searching through menus.

For example, if you wanted to draw a Circle you could type: **c <enter>**

You may turn the Command Line ON and OFF using Ctrl + 9.

DYNAMIC INPUT

If you choose to use Dynamic Input (DYN button on page 1-12) the command will be entered in the tooltip box beside the cursor.

Tooltip Box

The prompt will then appear.

Prompt

Specify center point for circle or ☒ 7.1024 12.3246

If you press the ↓ down arrow the options will appear below the prompt.

Specify center point for circle or ☒ 3.7915 11.6791
3P
2P
Ttr (tan tan radius)

Options

APPLICATION WINDOW DESCRIPTIONS....continued

DRAWING AREA
The Drawing Area is the large open area of the screen. This is where you will draw. Consider this your paper.
The color of this area can be changed using **Options / Display tab / Color**

UCS Icon
World Coordinate System (WCS) icon indicates the location of the Origin.
The Origin is where the coordinates X and Y originate.
The X and Y coordinates for the Origin is 0, 0.
This will be discussed more in future Lessons.

CURSOR
The Cursor is located within the Drawing Area. The movement of the pointing device, such as a mouse, controls the movement of the cursor. You will use the cursor to locate points, make selections and draw objects.
The size can be changed using **Options / Display tab / Crosshair Size**.

INFOCENTER
The InfoCenter is a tool to search for information. It is located in the upper right corner of the screen.

Help Box
The InfoCenter allows you to search for information by typing key words or a question in the "Help Box".

After typing press <enter> or click on "Search". AutoCAD searches it's help resources and displays links on a panel. You may click on any of the links to display the help topic, article or document.

Communication
The Communication panel displays links to information about product updates and announcements from Autodesk.

Favorites
Displays the Favorite panel to access saved topics.

Note:
To view an animated demonstration of the InfoCenter go to the Help menu described on the next page.

AUTOCAD'S HELP

This workbook was created to make your AutoCAD learning experience fun and uncomplicated.   To accomplish this I do not include every little detail about each command. I teach you the "meat and potatoes" of AutoCAD. If you would like to learn more about a command or system variable try AutoCAD's HELP .

How to open AutoCAD Help

1. Start a command.
2. Press the F1 key.

How to use AutoCAD Help
1. Select the "Index" tab.
2. Enter a keyword and press enter.
3. Help information will display in the right panel.

1   **2**   **3**

TOOLTIP HELP

Basic Tooltip
When you hover your cursor over a tool an initial Tooltip will appear telling you the name of the tool with a brief description.

Extended Tooltip
If you hover just a little longer a graphic display, directly from the Help system, will appear.

How to turn off Tooltips
After you become familiar with AutoCAD you will want to turn these off. Or you may just want to delay the extended Tooltips.

1. Type **options** on the command line and press <enter>.

2. Select the **Display** tab

3. Uncheck boxes

LEARNING OBJECTIVES

After completing this lesson, you will be able to:

1. Create and use a Template
2. Select a Command
3. Draw, Select and Erase objects
4. Start a New drawing
5. Open an Existing Drawing
6. Save, Backup and Recover a drawing
7. Exit AutoCAD

LESSON 2

CREATE A TEMPLATE

The first item on the learning agenda is **how to create a template file** from a drawing file. **This is important:** You will need this template to complete Lessons 2 through 8.

First you need to download a few files.

A. Go to the website: **www.shrockpublishing.com**

B. Select "**Downloadable Files**".

C. Save them to your "**Desktop**".

Now you will create a template. (This will be a very easy task.)

1. Start AutoCAD, if you haven't already.   (Refer to page 1-2)

2. Select the **OPEN** tool from the **Quick Access Toolbar**.   (Refer to page 1-9)

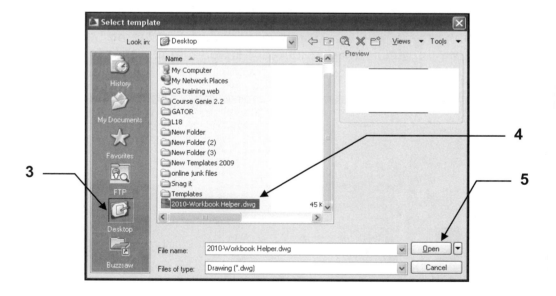

3. Select the **Desktop** directory

4. Select **2010-Workbook Helper.dwg**.

5. Select **OPEN**

Notice the 3 letter extension for a drawing file is .dwg

Continued on the next page…

CREATE A TEMPLATE....continued

6. Select the **"Application Menu"** ▾

7. Select **"Save As"** ▶

8. Select **"AutoCAD Drawing Template"**.

9. Type the new file name **2010-Workbook Helper** in the **File Name** box.
   Do not type the extension .dwt, AutoCAD will add it automatically.

10. Select the **Save** button.

Continued on the next page...

CREATE A TEMPLATE....continued

11. Type the description as shown below.

12. Select **OK** button.

11 —

12

Now you have a template to use for Lessons 2 through 8.

At the beginning of each exercise you will be instructed to start a NEW drawing using the **2010-Workbook Helper.dwt,**

Using a template as a master setup drawing is very good CAD management.

USING A TEMPLATE

The template that you created from the previous pages will be used for lessons 2 through 8. It will appear as a blank screen, but there are many variables that have been preset. This will allow you to start drawing immediately. You will learn how to set those variables before you complete this workbook, but for now you will concentrate on learning the AutoCAD commands and hopefully have some fun.

TO USE A TEMPLATE

1. Select the **NEW** tool from the **Quick Access Toolbar.**

2. Select **Drawing Template [*.dwt]** from the "Files of type: list.

3. Select the **2010-Workbook Helper.dwt** from the list of templates.

   Note: If you do not have this template, refer to page 2-1 for instructions.

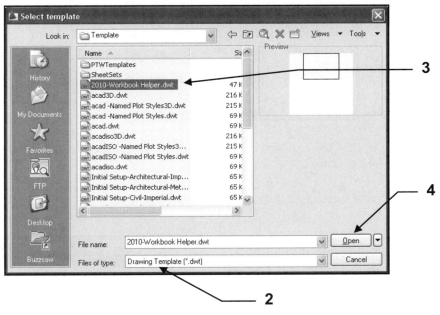

4. Select the **Open** button.

Note:
If you find that you have more than one drawing open you may find this confusing. I suggest that you refer to page 2-15 and follow the steps to change your AutoCAD system to only allow one drawing open at one time.

HOW TO SELECT A COMMAND

AutoCAD provides you with 2 different methods for selecting commands. One is <u>selecting a tool from the Ribbon</u>, the other is <u>typing the command</u>. Both methods will accomplish the same end result. You decide which method you prefer. An example of each are shown below.

Method 1: Selecting a tool from the Ribbon

1. First select a **tab** such as **Home**.

2. Locate the correct **Panel** such as **Draw**.

3. Select a **tool** such as **Circle**

1. Tab

3. Tool

2. Panel

Note:
If the tool includes multiple types it will have a down-arrow ▼ beside the tool.

If you select the down-arrow a sub-menu will appear.

Select the desired type such as **2-Point**.

The latest selection will then become the current displayed tool because AutoCAD assumes that you may need that tool again.

Sub-Menu

Continued on the next page...

HOW TO SELECT A COMMAND....continued

Method 2: Keyboard entry

You may type commands on the command line or in the Dynamic Input tooltip.
It depends on whether you have Dynamic Input On or Off.
Note: Additional information on page 1-13.

For example you could type the command **Circle** and then press <enter>.

If you have **Dynamic Input (DYN) ON**, the word **Circle** will appear within the Dynamic Input tooltip next to the cursor as shown below.

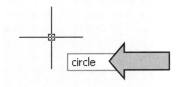

If you have **Dynamic Input OFF**, the word **Circle** will appear on the Command Lline as shown below.

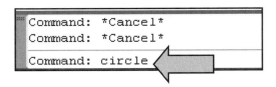

Dynamic Input can be turned **ON** or **OFF** by selecting the Dynamic Input button located on the Status Bar or pressing the **F12** key. (Refer to page 1-12)

Note:
Some commands have abbreviated entries such as **L** for **Line** or **C** for **Circle**. These abbreviations are called Aliases. You may find a listing of these aliases' in the acad.pgp or acadlt.pgp files. This list may also be customized.

DRAWING LINES

A **Line** can be **one segment** or a **series of connected segments.**
But each segment is an individual object.

One Segment   **Series of connected Segments**

Start the Line command using one of the following methods:

Ribbon = Home tab / Draw Panel /

Keyboard = L <enter>

Lines are drawn by specifying the locations for each endpoint.
Move the cursor to the location of the **"first"** endpoint (1) then press the left mouse
button. Move the cursor again to the **"next"** endpoint (2) and press the left mouse
button. Continue locating **"next"** endpoints until you want to stop.

There are 2 ways to **Stop drawing a line**:
1. Press <enter> key.
2. Press <Space Bar>

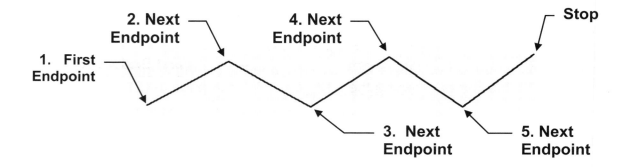

Helpful hint:
To quickly repeat the Line command, press the Spacebar.

Continued on the next page...

DRAWING LINES....continued

Horizontal and Vertical Lines

To draw a Line perfectly Horizontal or Vertical select the **Ortho** mode by selecting the **Ortho** button on the Status Bar or pressing the **F8** key.

 or

Try the following exercise:

1. Select the Line command.

2. Place the **First endpoint**

3. **Turn Ortho ON** by selecting the Ortho button or F8. (The "Ortho" button will change to blue when ON.)

4. Move the cursor to the right and press the left mouse button to place the **next endpoint.**   (The line should appear perfectly horizontal.)

5. Move the cursor down and press the left mouse button to place the next endpoint. (The line should appear perfectly vertical)

6. Now turn Ortho **OFF** by selecting the Ortho button. (The "Ortho" button will change to gray when OFF.)

7. Now move the cursor up and to the right at an upper ward angle. (The line should move freely now)

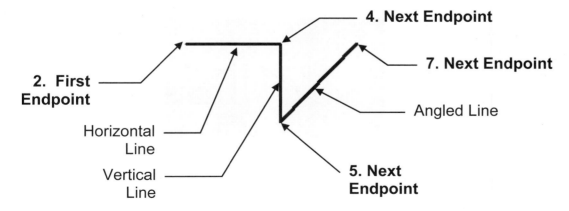

Ortho can be turned ON or OFF at any time while you are drawing. It can also be turned OFF temporarily by holding down the **Shift** key. Release the Shift key to resume.

Continued on the next page...

DRAWING LINES....continued

Closing Lines

If you have drawn 2 or more line segments, the <u>endpoint of the last line segment </u>can be connected automatically to the <u>first endpoint </u>using the **Close** option.

Try the following exercise:

1. Select the Line command.

2. Place the **First endpoint**

3. Place the next endpoint

4. Place the next endpoint

5. Type **C** <enter>

   Or

5. <u>Press the right mouse button </u>and select **Close** from the **Shortcut** menu.

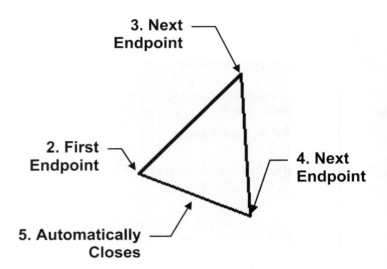

3. Next Endpoint

2. First Endpoint

4. Next Endpoint

5. Automatically Closes

What is the Shortcut Menu?
The <u>Shortcut menu </u>gives you quick access to command options.

If you have Dynamic Input <u>OFF</u>:
Press the right mouse button.
The shortcut menu will appear.
Select an option.

If you have Dynamic Input <u>ON</u>:
You may use the right mouse button or press the down arrow ↓ and the options will appear below the Dynamic Input prompt.
(Refer to page 1-13)

METHODS OF SELECTING OBJECTS

Most AutoCAD commands prompt you to "**select objects**". This means select the objects that you want the command to effect.

There are 2 methods. **Method 1. Pick**, is very easy and should be used if you have only 1 or 2 objects to select. **Method 2. Window**, is a little more difficult but once mastered it is extremely helpful and time saving. Practice the examples shown below.

Method 1. PICK :
First start a command such as ERASE. (Press E <enter>)
Next you will be prompted to "**Select Objects**", place the cursor (pick box) on the object. The object will highlight.
This appearance change is called "Rollover Highlighting".
This gives you a preview of which object will be selected.
Press the left mouse button to select the highlighted object.

Method 2. WINDOW: Crossing and Window

Crossing:
Place your cursor in the area <u>up</u> and to the <u>right</u> of the objects that you wish to select (**P1**) and press the left mouse button.
Then move the cursor down and to the left (**P2**) and press the left mouse button again.
(Note: The window will be *green* and outer line is *dashed*.)
Only the objects that this window **crosses** will be selected.

In the example on the right, all 3 circles have been selected because the Crossing Window crosses a portion of each.

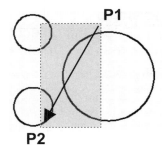

Window:
Place your cursor in the area <u>up</u> and to the <u>left</u> of the objects that you wish to select (**P1**) and press the left mouse button.
Then move the cursor <u>down</u> and to the <u>right</u> of the objects (**P2**) and press the left mouse button.
(Note: The window will be *blue* and outer line is *solid*.)
Only the objects that this window **completely enclosed** will be selected.

In the example on the right, only 2 circles have been selected.
(The large circle is **not** <u>completely enclosed</u>.)

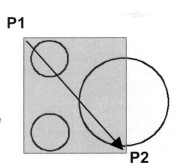

ERASE

There are 3 methods to erase (delete) objects from the drawing.
They all work equally well. You decide which one you prefer to use.

Method 1
Select the Erase command first and then select the objects.

Example:
1. Start the Erase command using one of the following:

    Ribbon = Home tab / Modify panel /

    Keyboard = E <enter>

2. Select objects: ***Pick one or more objects***
    Select Objects: ***Press <enter> and the objects selected will disappear.***

Method 2
Select the Objects first and then the Delete Key.

Example:
1. Select the object to be erased.
2. Press the Delete Key.

Method 3
Select the Objects first and then select Erase command from the Shortcut Menu.

Example:
1. Select the object to be erased.
2. Press the right Mouse button.
3. Select Erase from the Shortcut Menu.

Note: Very Important
If you want the erased objects to return, select the **Undo tool** from the **Quick Access Toolbar**. This will Undo the last command.
More about Undo and Redo later.

STARTING A NEW DRAWING

Starting A New Drawing means that you want to start with a previously created Template file. That is why I taught you "how to create a template" at the beginning of this lesson. You will use the **2010-Workbook Helper.dwt** template each time you are instructed to **Start a New Drawing**.

*Note: Do not use the **New** tool if you want to **open** an **existing drawing**.*
*Refer to page 2-14 to **OPEN** an existing drawing file.*

HOW TO START A NEW DRAWING

1. Select the **NEW** tool from the **Quick Access Toolbar**.

2. Select the **2010-Workbook Helper.dwt** from the list of templates.

   Note: If you do not have this template, refer to page 2-1 for instructions.

3. Select the **Open** button.

Note:
If you find that you have more than one drawing open you may find this confusing. I suggest that you refer to page 2-15 and follow the steps to change your AutoCAD system to only allow one drawing open at one time.

OPENING AN EXISTING DRAWING FILE

Opening an Existing Drawing File means that you would like to open on to the screen a drawing that has been previously created and saved. Usually you are opening it to continue working on it or you need to make some changes.

1. Select the **OPEN** tool on the **Quick Access Toolbar**.

2. Locate the Directory and Folder in which the file had previously been saved.

3. Select the File that you wish to OPEN.

4. A **Thumbnail Preview image** of the file selected is displayed in the Preview area.

5. Select the **Open** button.

Note:
If you find that you have more than one drawing open you may find this confusing.
I suggest that you refer to page 2-15 and follow the steps to change your AutoCAD system to only allow one drawing open at one time.

MULTIPLE OPEN FILES

AutoCAD allows you to have multiple drawing files open at the same time. If you select the **Application Menu** and select the **Open Documents** icon a list of all open drawings will be displayed.

Application Menu

Open Document icon

List of open drawings

This is very helpful to **experienced** AutoCAD users but can be confusing for users **NEW** to AutoCAD.

While using this workbook you might find it helpful if you <u>change a setting</u> to prevent opening more than one file at one time. This will only change AutoCAD and will have <u>no affect on any other software on your computer</u>.

1. Close all but one drawing file. You should only have one drawing open.
2. Type: sdi <enter>
3. Type: 1 <enter>

Now AutoCAD will restrict you to one drawing on the screen. If you choose to go back to multiple drawings repeat the steps above except enter "0" instead of "1".

Read-Only files

If the SDI setting is off, allowing multiple open files, and you open a drawing file that you <u>already have open</u> AutoCAD will display this warning.

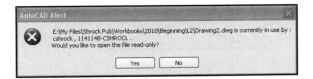

It is best if you select the **No** button. If you select the **Yes** button, AutoCAD will open a duplicate file as a **"Read-only"** file. A Read-only file should only be used as a reference file. It is not good practice to work on a Read-only file.

SAVING A DRAWING FILE

After you have completed a drawing, it is very important to save it. Learning how to save a drawing correctly is almost more important than making the drawing. If you can't save correctly, you will lose the drawing and hours of work.

There are 2 commands for saving a drawing: **Save** and **Save As**.
I prefer to use **Save As**.

The **Save As** command always pauses to allow you to choose where you want to store the file and what name to assign to the file. This may seem like a small thing, but it has saved me many times from saving a drawing on top of another drawing by mistake.

The **Save** command will automatically save the file either back to where you retrieved it or where you last saved a previous drawing. Neither may be the correct destination. And may replace a file with the same name. So play it safe, use **Save As** for now.

1. Select the **Saveas** command using one of the following:

Quick Access Toolbar =

Application Menu = Save As / AutoCad Drawing

Keyboard = Saveas <enter>

> Notice that I added the **Saveas** tool to the <u>Quick Access Toolbar.</u> If you would like to do this also, follow the easy instructions on page 1-9

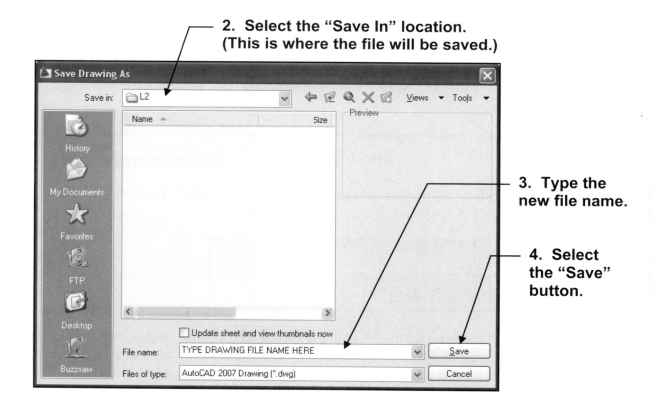

**2. Select the "Save In" location.
(This is where the file will be saved.)**

3. Type the new file name.

4. Select the "Save" button.

AUTOMATIC SAVE

AUTOMATIC SAVE

If you turn the automatic save option ON, your drawing is saved at specified time intervals. These temporary files are automatically deleted when a drawing closes normally. The default save time is every 10 minutes. You may change the save time Intervals and where you would prefer the Automatic Save files to be saved.

How to set the Automatic Save intervals

1. Type **options <enter>**

2. Select the **Open and Save** tab.

3. Enter the desired **minutes between saves**.

4. Select the **OK** button.

How to change the Automatic Save location

1. Type **options <enter>**

2. Select the **Files** tab

3. Locate the **Automatic Save File Location** and click on the + to display the **path**.

4. Double click on the path

5. Browse to locate the Automatic Save Location desired and highlight it.

6. Select **OK**.

(The browse box will disappear and the new location path should be displayed under the Automatic Save File Location heading)

7. Select **OK** to accept the change.

BACK UP FILES and RECOVER

BACK UP FILES

When you save a drawing file, Autocad creates a file with a **.dwg** extension.
For example, if you save a drawing as **12b**, Autocad saves it as **12b.dwg**. The next
time you save that same drawing, Autocad replaces the old with the new and renames
the old version **12b.bak**. The old version is now a back up file.
(Only 1 backup file is stored.)

How to open a back up file:
You can't open a **.bak** file.
It must first be renamed with a **.dwg** file extension.

How to view the list of back up files:
The backup files will be saved in the same location as the drawing file.
You must use Windows Explorer to locate the .bak files.

How to rename a back up file:
Right click on the file name.
Select "Rename".
Change the .bak extension to .dwg and press <enter>.

RECOVERING A DRAWING

In the event of a program failure or a power failure open files should be saved
automatically. (Refer to page 2-17)

When you attempt to re-open the drawing the **Drawing Recovery Manager** will
display a list of all drawing files that were open at the time of a program or system
failure. You can preview and open each .dwg or .bak file to choose which one should
be saved as the primary file.

EXITING AUTOCAD

To safely exit AutoCAD follow the instructions below.

1. Save all open drawings.

2. Start the **EXIT** procedure using one of the following.

   Ribbon = None

   Application Menu = | Exit AutoCAD |

   Keyboard = Exit <enter>

If any changes have been made to the drawing since the last **Save As**, the warning box shown below will appear asking if you want to *SAVE THE CHANGES*?

Select *YES, NO or CANCEL*.

EXERCISE 2A

INSTRUCTIONS:

1. Start a **NEW** file using **2010-Workbook Helper.dwt.**
2. **Draw** the objects below using:
   - A. **Line** command
   - B. **Ortho** (F8) **ON** when drawing Horizontal and Vertical lines.
   - C. **Ortho** (F8) **OFF** when drawing lines on an Angle.
   - D. Turn **Increment Snap** (F9) **ON**
   - E. Turn **Osnap** (F3) **OFF**
   - F. Turn **Grid** (F7) **ON**
   - G. Use the **Close** option
3. **Save** the drawing as: **EX2A**

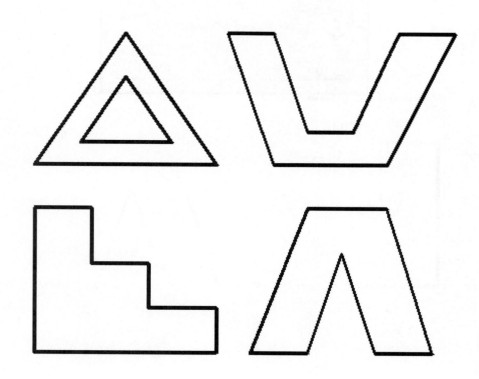

EXERCISE 2B

INSTRUCTIONS:

1. <u>**OPEN**</u> **EX2A**, if not already open.

2. **Erase** the missing Lines as shown.

      A. Turn **Osnap** (F3) **OFF** (It will be easier to move the cursor accurately)

3. **Save** the drawing as: **EX2B**

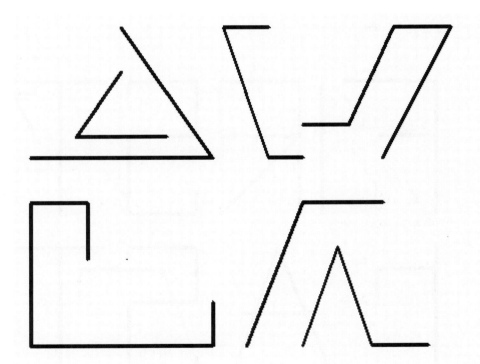

EXERCISE 2C

INSTRUCTIONS:

1. Start a **NEW** file using **2010-Workbook Helper.dwt.**
2. **Draw** the objects below using:
   - A. **Line** command
   - B. **Ortho** (F8) **ON** when drawing Horizontal and Vertical lines.
   - C. **Ortho** (F8) **OFF** when drawing lines on an Angle.
   - D. Turn **Increment Snap** (F9) **ON**
   - E. Turn **Osnap** (F3) **OFF**
   - F. Turn **Grid** (F7) ON
   - G. Use the **Shift Key** to toggle Ortho ON and OFF
3. **Save** the drawing as: **EX2C**

EXERCISE 2D

INSTRUCTIONS:

1. Start a **NEW** file using **2010-Workbook Helper.dwt.**

2. **Draw** the objects below using:

      A. **Line** command

      B. **Ortho** (F8) **ON** when drawing Horizontal and Vertical lines.

      C. **Ortho** (F8) **OFF** when drawing lines on an Angle.

      D. Turn **Increment Snap** (F9) **ON**

      E. Turn **Osnap** (F3) **OFF**

      F. Turn **Grid** (F7) ON

      G. Use the **Shift Key** to toggle Ortho ON and OFF

3. **Save** the drawing as: **EX2D**

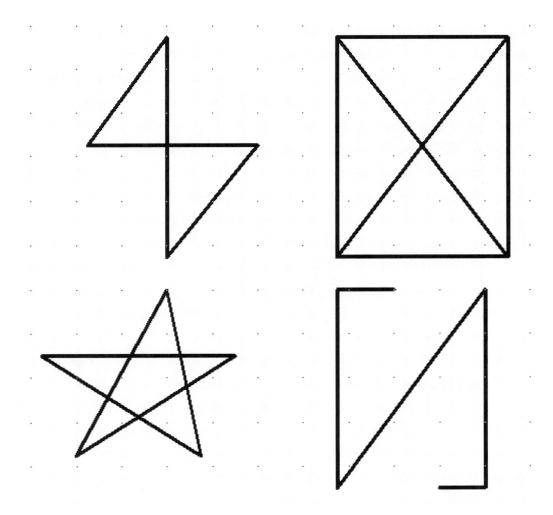

Notes:

LEARNING OBJECTIVES

After completing this lesson, you will be able to:

1. Create a Circle using 6 different methods
2. Create Rectangles with chamfers, fillets, width and rotation.
3. Understand how to Undo a command
4. Set Grids and Increment Snap
5. Draw using Layers
6. Control Layers
7. Create Layers
8. Assign Lineweights

LESSON 3

CIRCLE

There are 6 options to create a circle.

The default option is "**Center, radius**". (Probably because that is the most common method of creating a circle.)

We will try the **"Center, radius"** option first.

1. Start the **Circle** command by using one of the following:

   Ribbon = Home tab / Draw panel / ⊙ ▾

   Keyboard = C <enter>

2. The following will appear on the command line:
   Command: _circle Specify center point for circle or [3P/2P/Ttr (tan tan radius)]:

3. Locate the center point for the circle by moving the cursor to the desired location in the drawing area **(P1)** and press the left mouse button.

4. Now move the cursor away from the center point and you should see a circle forming.

5. When the circle is the size desired **(P2)**, press the left mouse button, or type the radius and then press <enter>.

Note: To use one of the other methods described below, first select the Circle command, then select one of the other Circle options.

Center, Radius: (Default option)
1. Specify the center (P1) location.

2. Specify the Radius (P2).

   (Define the Radius by moving the cursor or typing radius)

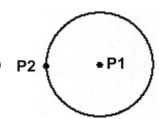

Center, Diameter:
1. Specify the center (P1) location.

2. Specify the Diameter (P2). (Define the Diameter by

   moving the cursor or typing Diameter)

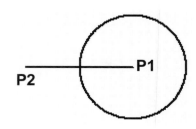

CIRCLE....continued

2 Points:
1. Select the 2 point option
2. Specify the 2 points (P1 and P2) that will determine the Diameter .

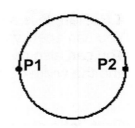

3 Points:
1. Select the 3 Point option
2. Specify the 3 points (P1, P2 and P3) on the circumference.
   The Circle will pass through all three point

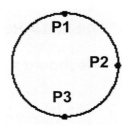

Tangent, Tangent, Radius:
1. Select the Tangent, Tangent, Radius option .
2. Select two objects (P1 and P2) for the Circle to be tangent to by placing the cursor on the object and pressing the left mouse button
3. Specify the radius.

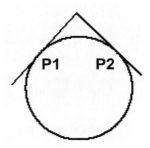

Tangent, Tangent, Tangent:
1. Select the Tangent, Tangent, Tangent option
2. Specify three objects (P1, P2 and P3) for the Circle to be tangent to by placing the cursor on each of the objects and pressing the left mouse button. (AutoCAD will calculate the diameter.)

RECTANGLE

A Rectangle is a closed rectangular shape. It is one object not 4 lines.
You can specify the length, width, area, and rotation options.
You can also control the type of corners on the rectangle—fillet, chamfer, or square and the width of the Line.

First, let's start with a simple Rectangle using the cursor to select the corners.

1. Start the **RECTANGLE** command by using one of the following:

    Ribbon = Home tab / Draw panel /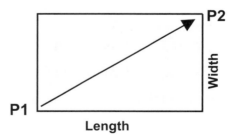

    Keyboard = REC <enter>

2. The following will appear on the command line:

    Command: _rectang
    Specify first corner point or [Chamfer/Elevation/Fillet/Thickness/Width]:

3. Specify the location of the first corner by moving the cursor to a location (**P1**) and then press the left mouse button.

    The following will appear on the command line:

    Specify other corner point or [Area / Dimensions / Rotation]:

4. Specify the location of the **diagonal** corner (**P2**) by moving the cursor diagonally away from the first corner (**P1**) and pressing the left mouse button.

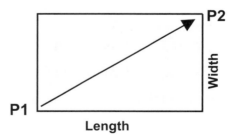

OR

4. Type **D <enter>**

    Specify length for rectangles <0.000>: *Type the desired length <enter>.*

    Specify width for rectangles <0.000>: *Type the desired width <enter>.*

    Specify other corner point or [Dimension]: *move the cursor up, down, right or left to specify where you want the second corner relative to the first corner and then press <enter> or press left mouse button.*

RECTANGLE....continued

OPTIONS: Chamfer, Fillet and Width

Note: the following options are available **Before** you place the **first corner** of the Rectangle.

CHAMFER

A chamfer is an angled corner. The Chamfer option automatically draws all 4 corners with chamfers simultaneously and all the same size. You must specify the distance for each side of the corner as distance 1 and distance 2.

Example: A Rectangle with dist1 = .50 and dist2 = .25

1. Select the Rectangle command
2. Type C <enter>
3. Enter .50 for the first distance
4. Enter .25 for the second distance
5. Place the first corner (P1)
6. Place the diagonal corner (P2)

FILLET

A fillet is a rounded corner. The fillet option automatically draws all 4 corners with fillets (all the same size). You must specify the radius for the rounded corners.

Example: A Rectangle with .50 radius corners.
1. Select the Rectangle command
2. Type F <enter>
3. Enter .50 for the radius.
4. Place the first corner (P1)
5. Place the diagonal corner (P2)

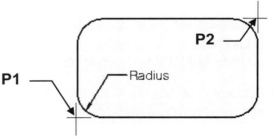

Note: You must set Chamfer and Fillet back to "0" before defining the width. Unless you want fat lines _and_ Chamfered or Filleted corners.

WIDTH

Defines the width of the rectangle lines.
Note: Do not confuse this with the "Dimensions" Length and Width.
Width makes the lines appear fatter.

Example: A Rectangle with a width of .50

1. Select the Rectangle command
2. Type W <enter>
3. Enter .50 for the width.
4. Place the first corner (P1)
5. Place the diagonal corner (P2)

RECTANGLE....continued

OPTIONS: Area and Rotation

Note: the following options are available **AFTER** you place the **first corner** of the Rectangle.

AREA

Creates a Rectangle using the AREA and either a LENGTH or a WIDTH. If the Chamfer or Fillet option is active, the area includes the effect of the chamfers or fillets on the corners of the rectangle.

Example: A Rectangle with an Area of 6 and a Length of 2.

1. Select the Rectangle command
2. Place the first corner (P1)
3. Type A <enter> for Area.
4. Enter 6 <enter> for the Area
5. Enter L <enter> for length option
6. Enter 2 <enter> for the length
(The width will automatically be calculated)

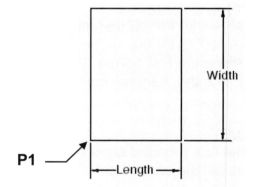

ROTATION

You may select the desired rotation angle <u>after</u> you place the first corner and <u>before</u> you place the second corner. The base point (pivot point) is the first corner.
<u>Note: All **new** rectangles within the drawing will also be rotated unless you reset the rotation to 0. This option will not effect rectangles already in the drawing.</u>

Example: A Rectangle with a rotation angle of 45 degrees.

1. Select the Rectangle command
2. Place the first corner (P1)
3. Type R <enter> for rotation.
4. Enter 45 <enter>
5. Place the diagonal corner (P2)

UNDO and REDO

The **UNDO** and **REDO** tools allow you to undo or redo previous commands.
For example, if you erase an object by mistake, you can UNDO the previous "erase"
command and the object will reappear. So don't panic if you do something wrong.
Just use the UNDO command to remove the previous commands.

The **Undo** and **Redo** tools are located in the **Quick Access Toolbar**.

Note:
You may UNDO commands used during a work session until you close the drawing.

How to use the <u>Undo</u> tool.

1. Start a new drawing.
2. Draw a line, circle and a rectangle.

Your drawing should look approximately like this.

3. Next Erase the Circle and the Rectangle.

   (The Circle and the Rectangle disappear.)

4. Select the UNDO arrow.

You have now deleted the ERASE command operation.
As a result the erased objects reappear.

How to use the <u>Redo</u> command:
Select the REDO arrow and the Circle and Rectangle will disappear again.

GRID and INCREMENT SNAP

GRID is the rectangular pattern of dots in the drawing area. The grid is only a drawing aid to assist you in aligning objects and visualizing the distances between them. The Grid <u>will not plot</u>. (Refer to page 1-12)

INCREMENT SNAP controls the movement of the cursor. If it is **OFF** the cursor will move smoothly. If it is **ON**, the cursor will jump in an *incremental* movement. (Refer to page 1-12)

The **DRAFTING SETTINGS** dialog box allows you to set **INCREMENT SNAP** and **GRID** spacing. You may change the Grid Spacing and Increment Snap at anytime while creating a drawing. The settings are only drawing aids to help you visualize the size of the drawing and control the movement of the cursor.

1. Select **DRAFTING SETTINGS** by using one of the following:

> **Keyboard = DS <enter>**

> **Status Bar = Right Click on SNAP or GRID button and select SETTINGS.**

2. The dialog box shown below will appear.

3 →

Make changes to the Snap and Grid settings here.

4

Open your template, confirm that these 3 boxes are not checked and resave your template.

3. Select the **"Snap and Grid"** tab.

4. Make your changes and select the **OK** button to save them.
   If you select the **CANCEL** button, your changes will **not** be saved.

LAYERS

A **LAYER** is like a transparency. Have you ever used an overhead light projector?
Remember those transparencies that are laid on top of the light projector? You could
stack multiple sheets but the projected image would have the appearance of one
document. Layers are basically the same. Multiple layers can be used within one
drawing.

The example, on the right, shows 3 layers.
One for annotations (text), one for dimensions and one
for objects.

HOW TO USE LAYERS
First you select the layer and then you draw the objects.
Always select the layer first and then draw the objects.

It is good "drawing management" to draw related objects on the same layer.
For example, in an architectural drawing, you would select layer "walls" and then draw
the floor plan.
Then you would select the layer "Electrical" and draw the electrical objects.
Then you would select the layer "Plumbing" and draw the plumbing objects.
Each layer can then be controlled independently.
If a layer is <u>Frozen</u>, it is <u>not visible</u>. When you <u>Thaw</u> the layer it becomes <u>visible</u> again.
(Refer to the following pages for detailed instructions for controlling layers.)

HOW TO SELECT A LAYER

1. Go to **Ribbon = Home tab / Layers panel**

2. Select the drop down arrow ▼

3. Highlight the desired layer and press the
   left mouse button.

CONTROLLING LAYERS

The following controls can be accessed using the Layer drop down arrow ▼.

ON or OFF
If a layer is **ON** it is **visible**. If a layer is **OFF** it is **not visible**.
Only layers that are **ON** can be **edited** or **plotted**.

FREEZE or THAW
Freeze and **Thaw** are very similar to On and Off. A Frozen layer is <u>not visible</u> and a Thawed layer <u>is visible</u>. Only thawed layers can be edited or plotted.

Additionally:
a. Objects on a Frozen layer **cannot** be accidentally erased
b. When working with large and complex drawings, freezing saves time because frozen layers are not **regenerated** when you zoom in and out.

LOCK or UNLOCK
Locked layers are visible but <u>cannot be edited</u>.
They are visible so they **will** be plotted.

CONTROLLING LAYERS....continued

To access the following options you must use the **Layer Properties Manager**.
You may also access the options listed on the previous page within this dialog box.

To open the **Layer Properties Manager** use one of the following.

Ribbon = Home tab / Layers panel /

Keyboard = LA <enter>

New Layer
Page 3-15

Delete Layer:
See below

Set Layer Current

Lineweights:
Refer to pages
3-13 and 3-14

Current
Layer

Color:
Page 3-12

Linetype:
Page 3-16

Not Plottable:
See below

HOW TO DELETE AN EXISTING LAYER

1. Highlight the layer name to be deleted.
2. Select the **Delete Layer** tool.

Or

1. Highlight the layer name to be deleted.
2. Right click and select **Delete Layer**

PLOT or NOT PLOTTABLE

This tool prevents a layer from plotting even though it is visible within the
Drawing Area.
A <u>Not Plottable</u> layer will not be displayed when using <u>Plot Preview</u>.
If the Plot tool has a slash the layer will not plot.

LAYER COLOR

Color is not merely to make a pretty display on the screen. Layer colors can help define objects. For example, you may assign color Green for all doors. Then, at a glance, you could identify the door and the layer by their color.

Here are some additional things to consider when selecting the colors for your layers.

Consider how the colors will appear on the paper.
(Pastels do not display well on white paper.)

Consider how the colors will appear on the screen.
(Yellow appears well on a black background but not on white.

How to change the color of a layer.

2. **Select the Color Swatch or word**

1. **Select the layer that you want to change.**

3. **Select the color from the Index or primary color strip. (The color name or number will appear in the color box.)**

4. **Select OK button**

5. **The color selected will appear on the layer line.**

LINEWEIGHTS

A **Lineweight** means "**how heavy or thin is the object line**".

It is "<u>good drawing management</u>" to establish a contrast in the lineweights of entities.

In the example below the rectangle has a heavier lineweight than the dimensions. The contrast in lineweights makes it easier to distinguish between entities.

LINEWEIGHT SETTINGS
Lineweights are plotted with the exact width of the lineweight assigned.
But you may adjust how they are <u>displayed on the screen</u>. (Refer to #3 below)

IMPORTANT: Before assigning lineweights you should first select the **Units** and **Display Scale** as shown below.

1. Select the **Lineweight Settings** box using one of the following:

> **Ribbon = Not discussed yet**
>
> **Keyboard = lweight <enter>**
>
> **Status Bar = Rt. Click on LWT button and select Settings.**

1. Select Inches or MM

2. Select Display Lineweight box

3. Click and drag to adjust the "Displayed Scale" of the lineweight. (Note: This will not effect the width when plotted)

4. Select OK button

NOTE: These settings will be saved to the computer not the drawing and will remain until you change them.

ASSIGNING LINEWEIGHTS

Note: Before assigning **Lineweights** to Layers make sure your **Lineweight settings** (Units and Display scale) are correct. Refer to the previous page.

ASSIGNING LINEWEIGHTS TO LAYERS

1. Select the Layer Properties Manager using one of the following:

    Ribbon = Home tab / Modify panel /

    Keyboard = LA <enter>

2. Highlight a Layer (Click on the name)

3. Click on the Lineweight for that layer.

4. Scroll and select a lineweight from the list.

5. Select the **OK** button.

Note: Lineweight selections will be saved within the **current** drawing and will not effect any other drawing.

CREATING NEW LAYERS

Using layers is an important part of managing and controlling your drawing. It is better to have too many layers than too few. You should draw like objects on the same layer. For example, place all doors on the layer "door" or centerlines on the layer "centerline".

When you create a new layer you will assign a **name, color, linetype, lineweight** and whether or not it should **plot**.

1. Select the Layer command using one of the following:

   Ribbon = Home tab / Layers panel /

   Keyboard = LA <enter>

2. New Layer tool

3

4. Features

2. Select the **New Layer tool** and a new layer will appear.

3. Type the new layer **name** and press <enter>

4. Select any of the **features** and a dialog box will appear.

Features:
Refer to the previous pages for controlling and selecting color and lineweights.

LOADING and SELECTING LAYER LINETYPES

In an effort to conserve data within a drawing file, AutoCAD automatically loads only one linetype called "**continuous**". If you would like to use other **linetypes**, such as "dashed" or "fenceline", you must **Load** them into the drawing as follows:

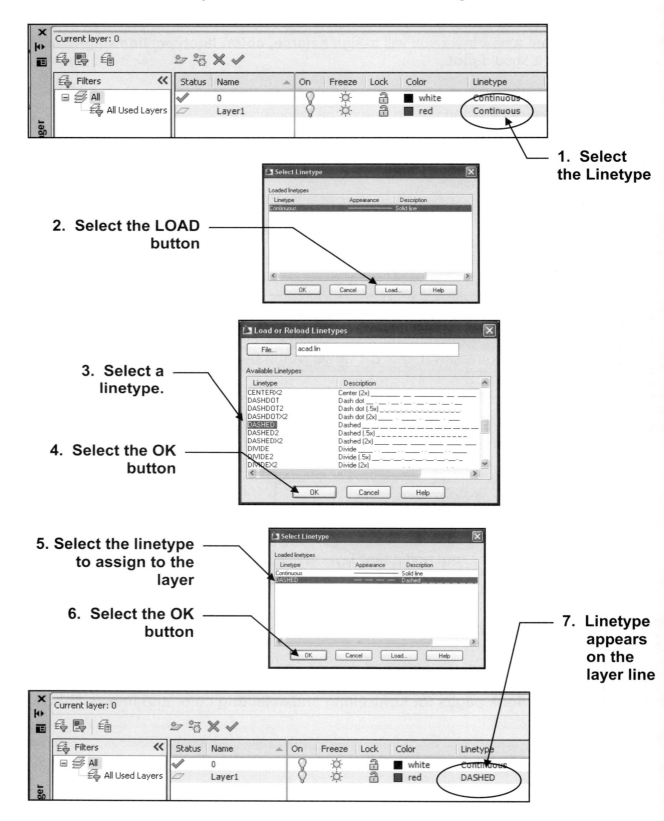

1. Select the Linetype

2. Select the LOAD button

3. Select a linetype.

4. Select the OK button

5. Select the linetype to assign to the layer

6. Select the OK button

7. Linetype appears on the layer line

EXERCISE 3A

INSTRUCTIONS:

1. Start a **NEW** file using **2010-Workbook Helper.dwt.**

2. **Draw** the **LINES** below using:

    A. **Line** command

    B. **Ortho** (F8) **ON**.

    C. Turn **Increment Snap** (F9) **ON**

3. Select the appropriate layer, then select the Line command, then draw a line.

4. Select the appropriate layer, then select the Line command, then draw a line.

5. Select the appropriate layer, then select the Line command, then draw a line.

    Are you noticing a pattern here?

6. **Save** the drawing as: **EX3A**

Do not enter text.

You will learn Text soon.

Label	
BORDERLINE	————————————————
CENTERLINE	——— — ——— — ——— — ———
DIMENSION	————————————————
DOOR	————————————————
HATCH	————————————————
HIDDEN LINE	– – – – – – – – – – – –
OBJECT LINE	————————————————
PHANTOM LINE	——— — — ——— — — ———
PLANTS	————————————————
ROOF	————————————————
SECTION LINE	——— — — ——— — — ———
TEXT	————————————————
WALL	————————————————
WINDOW	————————————————

EXERCISE 3B

INSTRUCTIONS:

1. Start a **NEW** file using **2010-Workbook Helper.dwt.**
2. Change the **Increment Snap** to **.20** and **Grid spacing** to **.40**

    If you have the Snap and Grids set correctly it will be easy.

3. Draw the objects below using **layer Object Line**.
4. **Do not dimension**, you will learn dimensioning soon.
5. **Save** the drawing as: **EX3B**

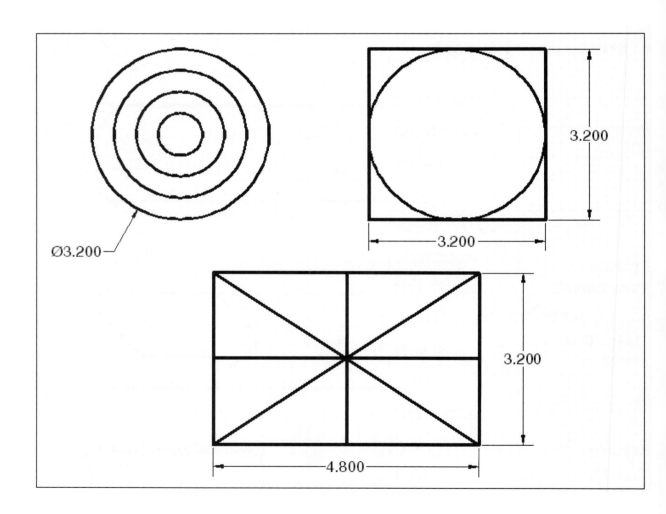

EXERCISE 3C

INSTRUCTIONS:

1. Start a **NEW** file using **2010-Workbook Helper.dwt.**
2. **Draw** the **Rectangles** below using the options:

   Dimension, Chamfer, Fillet, Width and Rotation
3. Use layer **Object Line.**
4. **Save** the drawing as: **EX3C**

Rectangle Dimensions: Length = 3   Width = 2

       Chamfer = Dist1 = .50   Dist2 = .50

Rectangle Dimensions: Length = 3 Width = 2

       Fillet = .50

Rectangle Dimensions: Length = 3 Width = 2

       Rotation = 45

       Chamfer = 0

       Fillet = 0

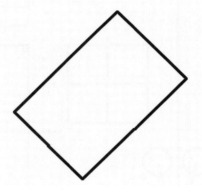

Rectangle Dimensions: Length = 3 Width = 2

       Width = .200

       Rotation = 0

       Chamfer = 0

       Fillet = 0

EXERCISE 3D

INSTRUCTIONS:

1. Start a **NEW** file using **2010-Workbook Helper.dwt.**

2. **Important, important: Draw** the objects below using the following Layers:

    Roof, Wall, Window, Door and Plants

3. You can change Snap and Grid settings to whatever you like.

4. You decide when to turn Ortho and Snap ON or OFF.

5. All objects must be place accurately. All lines must intersect exactly.

6. **Save** the drawing as: **EX3D**

EXERCISE 3E

INSTRUCTIONS:

1. Open **EX3D.**

2. Make layer **Object** "<u>current</u>"

(I am having you do that because you can't Freeze or Lock a layer that is <u>current </u>and I want to make sure that none of the layers that you used in the drawing is <u>current</u>.)

3. **Freeze** the following layers: <u>Window</u> and <u>Plants </u>(Do not use erase)

4. **Lock** layer <u>Roof</u>

5. Try to erase any of the roof lines.

You can't because the Roof layer is <u>Locked</u>.

6. **Save** the drawing as: **EX3E**

EXERCISE 3F

INSTRUCTIONS:

1. Start a new file using **Acad.dwt** (**do not select** Acad3D.dwt)

   (Located in the Template list below your 2010-Workbook Helper template)

2. Load Linetypes **Zigzag, Gas Line and Dashed**

3. Create 3 layers named **Fence, Gas and Invisible.**

4. **Draw** the **LINES** below using the new layers:

   A. Select the layer

   B. Select the **Line** command

   C. Draw 5 inche long line.

   D. **Ortho** (F8) **ON**.

5. **Save** the drawing as: **EX3F**

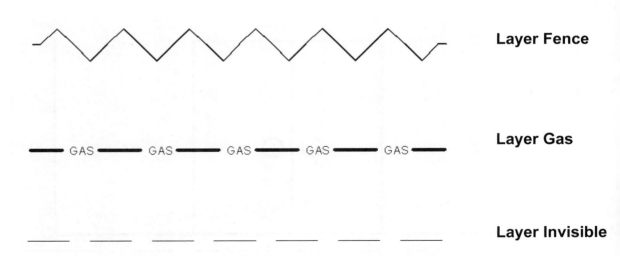

Layer Fence

Layer Gas

Layer Invisible

LEARNING OBJECTIVES

After completing this lesson, you will be able to:

1. Understand Object Snap
2. Use Running Object Snap
3. Use the Zoom options to view the drawing
4. Change the Drawing Limits
5. Select the Units of Measurement and Precision

LESSON 4

OBJECT SNAP

In Lesson 3 you learned about <u>Increment Snap.</u>   Increment Snap enables the cursor to move in an incremental movement. So you could say your cursor is "snapping to increments" preset by you.

Now you will learn about <u>Object Snap</u>. If Increment Snap snaps to increments, what do you think Object Snap snaps to? That's right; "objects". Object snap enables you to snap to "objects" in very specific and accurate locations on the objects.
For example, the endpoint of a line or the center of a circle.

How to select from the Object Snap Menu

1. You must select a command, such as LINE, before you can select Object Snap.

2. While holding down the <u>shift key</u>, <u>press the right mouse button</u>   The menu shown below should appear.

3. Highlight and press left mouse button to select.

The following object snaps will be discussed in this Lesson:
Endpoint, Midpoint, Intersection, Center, Quadrant, Perpendicular and Tangent.
Refer to their descriptions on the next page.
The remaining will be discussed in future lessons.

OBJECT SNAP....continued

Object Snap Definitions
Object snap is used when AutoCAD prompts you to place an object. Object snap allows you to place objects very accurately.

A step by step example of "How to use object snap" is shown on the next page.

Note: You may type the **3 bold letters** shown rather than select from the menu.

 ENDpoint — Snaps to the closest endpoint of a Line, Arc or polygon segment. Place the cursor on the object close to the end and the cursor will snap like a magnet to the end of the line.

 MIDpoint — Snaps to the middle of a Line, Arc or Polygon segment. Place the cursor anywhere on the object and the cursor will snap like a magnet to the midpoint of the line.

 INTersection — Snaps to the intersections of any two objects. Place the cursor directly on top of the intersection or select one object and then the other and Autocad will locate the intersection.

 CENter — Snaps to the center of an Arc, Circle or Donut. Place the cursor on the object, or at the approximate center location and the cursor will snap like a magnet to the center.

 QUAdrant — Snaps to a 12:00, 3:00, 6:00 or 9:00 o'clock location on a circle or ellipse. Place the cursor on the circle near the desired quadrant location and the cursor will snap to the closest quadrant.

 PERpendicular — Snaps to a point perpendicular to the object selected. Place the cursor anywhere on the object then pull the cursor away from object and press the left mouse button.

 TANgent — Calculates the tangent point of an Arc or Circle. Place the cursor on the object as near as possible to the expected tangent point.

How to use OBJECT SNAP

The following is an example of attaching a line segment to previously drawn vertical lines. The new line will start from the upper endpoint, to the midpoint, to the lower endpoint.

1.   Turn Off   **SNAP** and **OSNAP** on the Status Bar.

2.   Select the **Line** command.

3.   Draw two vertical lines as shown below

4.   Select the **Line** command again.

5.   Hold the shift key down and press the right mouse button.

6.   Select the Object snap **Endpoint** from the object snap menu.

7.   Place the cursor close to the upper endpoint
     of the left hand line.

     The cursor should snap to the end of the line like a magnet.
     A little red square and a "Endpoint" tooltip are displayed.

8.   Press the left mouse button to attach the new line to the upper endpoint of the previously drawn vertical line. (Do not end the Line command yet.)

Continued on the next page...

How to use OBJECT SNAP....continued

9. Now hold the <u>shift key down</u> and <u>press the right mouse button</u> and select the **Midpoint** object snap option.

10. Move the cursor to approximately the middle of the right hand vertical line.

<u>The cursor should snap to the midpoint of the line like a magnet .</u>
<u>A little red triangle with a "Midpoint" tool tip are displayed.</u>

11. Press the left mouse button to <u>attach</u> the new line to the midpoint of the previously drawn vertical line. (Do not end the Line command yet.)

12. Now hold the <u>shift key down</u> and <u>press the right mouse button</u> and select the object snap **Endpoint** again.

13. Move the cursor close to the lower endpoint of the left hand vertical line.

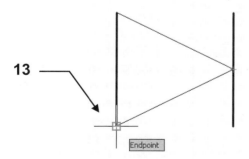

<u>The cursor should snap to the end of the line like a magnet.</u>
<u>A little red square and a tooltip are displayed.</u>

14. Press the left mouse button to <u>attach</u> the new line to the endpoint of the previously drawn vertical line.

15. Stop the Line command and disconnect by pressing **<enter>**.

RUNNING OBJECT SNAP

Selecting Object Snap is not difficult but AutoCAD has provided you with an additional method to increase your efficiency by allowing you to preset select frequently used object snap options. This method is called **RUNNING OBJECT SNAP.**

When **Running Object Snap** is active the cursor will automatically snap to any preset object snap locations thus eliminating the necessity of invoking the object snap menu for each locations.

First you must set the running object snaps and next you must turn ON the Running Object Snap option.

SETTING RUNNING OBJECT SNAP
1. Select the **Running Object Snap** dialog box using one of the following:

   Keyboard = DS <enter>

   Status Bar = Right Click on OSNAP button and select SETTINGS.

2. Select the **Object Snap** tab.

3. Select the Object Snaps desired.
   (In the example below object snap _Endpoint, Midpoint_ and _Intersection_ have been selected.

2

Note:
Try not to select more than 3 or 4 at one time.
If you select too many the cursor will flit around trying to snap to multiple snap locations. And possibly snap to the wrong location. You will lose control and it will confuse you.

4

4. Select the **OK** button.

5. Turn ON the **OSNAP** button on the Status Bar.

5

| SNAP | GRID | ORTHO | POLAR | OSNAP | OTRACK | DUCS | DYN | LWT | QP |

ZOOM

The **ZOOM** command is used to move closer to or farther away from an object. This is called Zooming In and Out.

1. Select the Zoom command by using the following:

   Ribbon = View tab / Navigate panel

2. Select the ▼ down arrow to display all of the selections.

The following are descriptions of the most commonly used zoom tools.

EXTENTS = Displays all objects in the drawing file, even objects outside of the drawing limits.

WINDOW = zoom in on an area by specifying a window around the area.

PREVIOUS = returns the screen to the previous display. (Limited to 10)

REAL TIME = Interactive Zoom. You can zoom in or out by moving the cursor vertically up or down while pressing the left mouse button.
To stop, press the Esc key.

All = Zooms to drawing limits or Extents, whichever is greater.

IN or OUT = moves in 2X or out 2X

You may also select the Zoom commands using one of the following:

Right Click and select Zoom from the Short cut menu.

Keyboard = Z <enter> Select from the options listed..

ZOOM....continued

How to use ZOOM / WINDOW

1. Select Zoom / Window (Refer to previous page)

2. Draw a window around the objects you want to enlarge.
    (Drawing a "window" is the similar to drawing a rectangle. It requires a first corner and then diagonal corner)

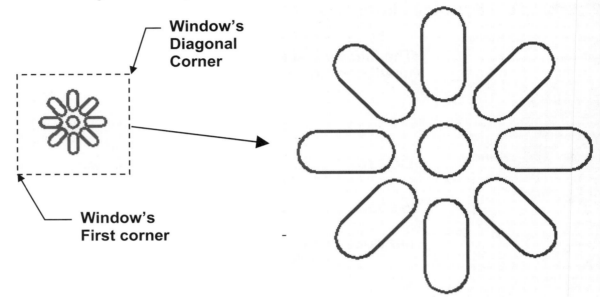

Window's Diagonal Corner

Window's First corner

<ins>Magnified to this view</ins>
**Note: the objects have been magnified.
But the size has not changed.**

How to return to Original View

1. Type: **Z** <enter> **A** <enter> (This is a shortcut for Zoom / All)

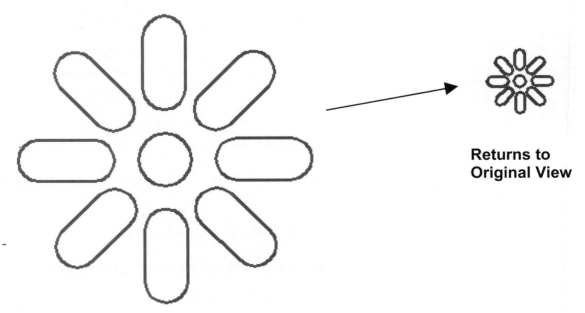

Returns to Original View

DRAWING SETUP

When drawing with a computer, you must "set up your drawing area" just as you would on your drawing board if you were drawing with pencil and paper. You must decide what size your paper you will need, what Units of measurement you will use (feet and inches or decimals, etc) and how precise you need to be. In CAD these decisions are called "Setting the **Drawing Limits, Units** and **Precision".**

AutoCAD 2010 also has a new feature "Initial Setup" that allows you to select an Industry, add task-based tools and specify the drawing template you want to use when creating a new drawing. See Help Menu for detailed instructions.

DRAWING LIMITS

Consider the drawing limits as the size of the paper you will be drawing on.
You will first be asked to define where the lower left corner should be placed, then the upper right corner, similar to drawing a Rectangle. An 11 x 8.5 piece of paper would have a **lower left corner** of 0,0 and an **upper right corner** of 11, 8.5. *(11 is the horizontal measurement X-axis and 8.5 is the vertical measurement Y-axis.)*

HOW TO SET THE DRAWING LIMITS

1. Select the **DRAWING LIMITS** command by typing: **Limits <enter>**

2. The following will appear on the command line:

   Command: '_limits
   Reset Model space limits:
   Specify lower left corner or [ON/OFF] <0.000,0.000>:

   **Note: Limits default setting is ON. AutoCAD will remind you when you are drawing outside of the Limits. If you change the limits to Off you may draw outside of the Limits.
   (See page 4-10)**

3. Type the X,Y coordinates **0, 0 <enter>** for the lower left corner location of your piece of paper .

4. The command line should now read:

   Specify upper right corner <11.000, 8.5>:

5. Type the X,Y coordinates **11, 8.5 <enter>** for the upper right corner of your piece of paper .

6. **This next step is very important:**

   Type **Z <enter> A<enter>** to make the screen display the new drawing limits.
   (This is the shortcut for Zoom / All. Refer to page 4-7)

DRAWING SETUP continued....

Grids within Limits

If you have your **Grid behavior** setting **Display grid beyond Limits** turned Off (no check mark) the grids will only be displayed within the Limits that you set.

**Off
(no check mark)**

**Grids displayed within
Limits only.**

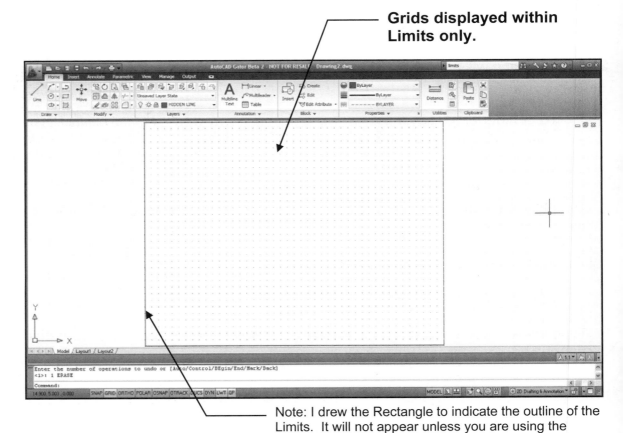

Note: I drew the Rectangle to indicate the outline of the Limits. It will not appear unless you are using the 2010-Workbook Helper template.

DRAWING SETUP continued....

Grids beyond Limits

If you have your **Grid behavior** setting **Display grid beyond Limits** turned **ON** (check mark) the grids will be displayed beyond the Limits that you set.

ON
(check mark)

Grids displayed beyond Limits.

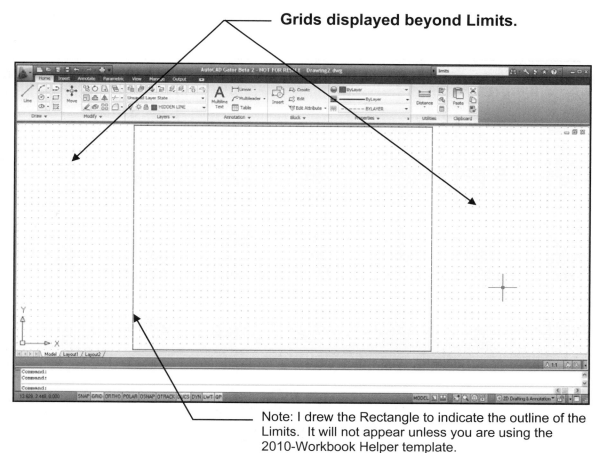

Note: I drew the Rectangle to indicate the outline of the Limits. It will not appear unless you are using the 2010-Workbook Helper template.

DRAWING SETUP....continued

UNITS AND PRECISION

You now need to select what unit of measurement with which you want to work.
Such as: Decimal (0.000) or Architectural (0'-0").
Next you should select how precise you want the measurements. For example, do you
want the measurement rounded off to a 3 place decimal or the nearest 1/8".

HOW TO SET THE UNITS AND PRECISION.

1. Select the **UNITS** command using one of the following:

    Application Menu = Drawing / Units

    Keyboard = Units <enter>

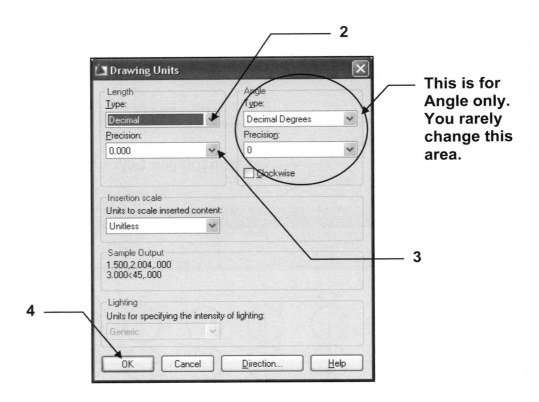

2. **Type:** Select the down arrow and select : **Decimals** or **Architectural**.

3. **Precision:** Select the down arrow and select the appropriate **Precision**
    associated with the "type".
    Examples: 0.000 for Decimals or 1/16" for Architectural.

4. Select the **OK** button to save your selections.

    Easy, yes?

EXERCISE 4A

INSTRUCTIONS:

1. Start a **NEW** file using **2010-Workbook Helper.dwt.**

2. Set **Units** and **Precision:**

   Units = Fractional   **Precision =** 1/2"

3. Set **Drawing Limits:**

   Lower Left corner = 0,0   Upper Right Corner = 20, 15

4. Make sure you use **Zoom / All** after setting Drawing Limits

5. Erase the Rectangle that appears with the template, it will appear too small.

6. Turn **OFF** the **Snap** and **Ortho.**

   (Your cursor should move freely)

7. Draw the objects shown below using:

   Circle, center radius and Line (Use Layer = Object Line)

   Object snap = Center and Tangent

8. **Save** the drawing as: **EX4A**

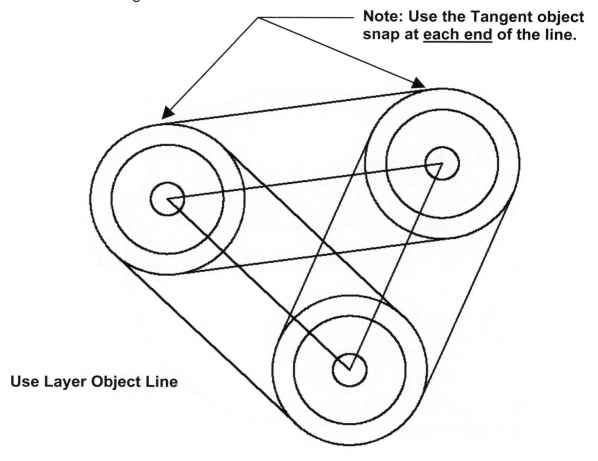

Note: Use the Tangent object snap at <u>each end</u> of the line.

Use Layer Object Line

EXERCISE 4B

INSTRUCTIONS:

1. Start a **NEW** file using **2010-Workbook Helper.dwt.**

2. Set **Units** and **Precision:**

   Units = Fractional   **Precision =** 1/4"

3. Set **Drawing Limits:**

   Lower Left corner = 0,0   Upper Right Corner = 12, 9

4. Make sure you use **Zoom / All** after setting Drawing Limits

5. Turn **OFF** the **Snap** and **Ortho.**

   (Your cursor should move freely)

6. Draw the objects shown below using:

   Circle, center radius (Use Layer = Object Line)

   Line (Use Layer = Hidden Line)

   Object snap = Quadrant

7. **Save** the drawing as: **EX4B**

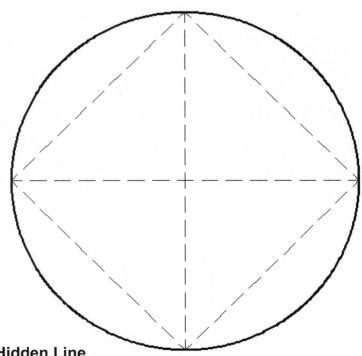

Use Layers:
Object Line and Hidden Line

EXERCISE 4C

INSTRUCTIONS:

1. Start a **NEW** file using **2010-Workbook Helper.dwt.**

2. Set **Units** and **Precision:**

> **Units =** Architectural   **Precision =** 1/2"

> **Note: A warning may appear asking you if you "are you sure you want to change the units?   Select OK button.**

3. Set **Drawing Limits:**

> Lower Left corner = 0,0   Upper Right Corner = 25, 20

4. Make sure you use **Zoom / All** after setting Drawing Limits

5. Erase the Rectangle that appears with the template

6. Turn **OFF** the **Snap** and **Ortho.**

> (Your cursor should move freely)

7. Draw the objects shown below using:

> Line (Use Layer = Object Line)

> Object snap = Perpendicular

8. **Save** the drawing as: **EX4C**

Start by drawing this line, then use Object Snap Perpendicular to draw the other lines.

Use Layer = Object Line

EXERCISE 4D

INSTRUCTIONS:

1. Start a **NEW** file using **2010-Workbook Helper.dwt.**

2. Set **Units** and **Precision:**

   Units = Decimals   **Precision =** 0.00

Note: A warning may appear asking you if you "are you sure you want to change the units?   Select OK button.

3. Set **Drawing Limits:**

   Lower Left corner = 0,0   Upper Right Corner = 12,9

4. Make sure you use **Zoom / All** after setting Drawing Limits

5. Turn **OFF Snap** and **Ortho.**

   (Your cursor should move freely)

6. Draw the objects shown below using:

   Line (Use Layer = Object Line)

   Object snap = Midpoint

7. **Save** the drawing as: **EX4D**

Start here

Use Layer = Object Line

EXERCISE 4E

INSTRUCTIONS:

1. Start a **NEW** file using **2010-Workbook Helper.dwt.**

2. Turn **OFF Snap.**

3. Turn **ON Ortho** for the Horizontal lines only.

4. Draw the Horizontal lines first with Ortho On

5. Turn Ortho Off and draw the angled lines using Object Snap Endpoint.

6. Draw the objects shown below using:

    Line (Use Layer = Object Line)

    Object snap = Endpoint

7. **Save** the drawing as: **EX4E**

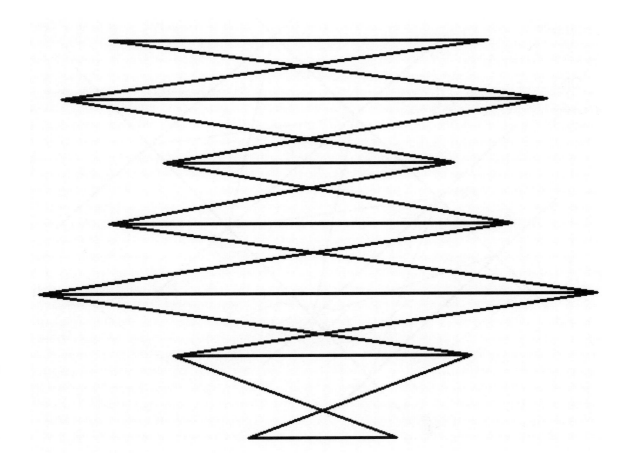

Use Layer = Object Line

EXERCISE 4F

INSTRUCTIONS:

1. Start a **NEW** file using **2010-Workbook Helper.dwt.**

2. Turn **OFF Snap** and **Ortho.**

3. Draw the objects shown below using:

    Line (Use Layer = Object Line)

    Object snap = Intersection

4. **Save** the drawing as: **EX4F**

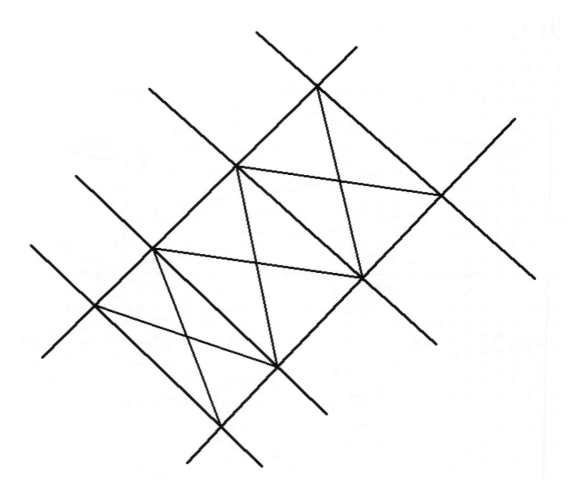

Use Layer = Object Line

EXERCISE 4G

INSTRUCTIONS:

1. Start a **NEW** file using **2010-Workbook Helper.dwt.**

2. Turn **OFF** **Snap** and **Ortho.**

3. Draw the 2 Circles on layer Object with the following Radii : 1.5 and 3.5.

    (Use <u>Object Snap: Center</u> so all circles will have the same center)

4. Draw the Lines using **Layers**: <u>Object Line</u> and <u>Centerline</u>

5. Use **Object Snap** <u>Quadrant</u> and <u>Tangent</u>

6. **Save** the drawing as: **EX4G**

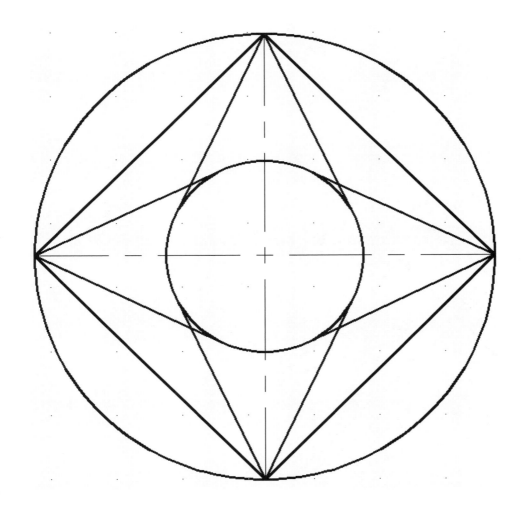

Use Layers = Object Line and Centerline

Notes:

LEARNING OBJECTIVES

After completing this lesson, you will be able to:

1. Draw an Inscribed or Circumscribed Polygon
2. Create an Ellipse using two different methods
3. Define an Elliptical Arc
4. Create Donuts
5. Define a Location with a Point
6. Select various Point Styles
7. Use 3 new Object Snap modes

LESSON 5

POLYGON

A polygon is an object with multiple edges (flat sides) of equal length. You may specify from 3 to 1024 sides. A polygon appears to be multiple lines but in fact it is one object. You can specify the <u>center and a radius</u> or the <u>edge length</u>. The <u>radius</u> size can be specified <u>Inscribed)</u> or <u>Circumscribed</u>.

CENTER, RADIUS METHOD
1. Select the **Polygon** command using one of the following:

   Ribbon = Home tab / Draw panel ▾ /

   Keyboard = POL <enter>

2. The following prompts will appear on the command line:

   _polygon Enter number of sides <4>: *type number of sides <enter>*
   Specify center of polygon or [Edge]: *specify the center location (P1)*
   Enter an option [Inscribed in circle/Circumscribed about circle]<I>:*type I or C<enter>*
   Specify radius of circle: *type radius or locate with cursor. (P2)*

Note:
The dashed circle is shown only as a reference to help you visualize the difference between Inscribed and Circumscribed. Notice that the radius is the same (2") but the Polygons are different sizes. Selecting Inscribed or Circumscribed is important.

INSCRIBED

CIRCUMSCRIBED

EDGE METHOD
1. Select the **Polygon** command using one of the options shown above.

2. The following prompts will appear on the command line:

   _polygon Enter number of sides <4>: *type number of sides <enter>*
   Specify center of polygon or [Edge]: *type E <enter>*
   Specify first endpoint of edge: *place first endpoint of edge (P1)*
   Specify second endpoint of edge: *place second endpoint of edge (P2)*

ELLIPSE

There are 3 methods to draw an Ellipse. You may (1) specify 3 points of the axes, (2) define the center point and the axis points or (3) define an elliptical Arc.
The following 3 pages illustrates each of the methods.

AXIS END METHOD

1. Select the **ELLIPSE** command using one of the following:

     Ribbon = Home tab / Draw panel /

     Keyboard = EL <enter>

2. The following prompts will appear on the command line:

     Command: _ellipse
     Specify axis endpoint of ellipse or [Arc/Center]: ***place the first point of either the major or minor axis (P1).***
     Specify other endpoint of axis: ***place the other point of the first axis (P2)***
     Specify distance to other axis or [Rotation]: ***place the point perpendicular to the first axis (P3).***

 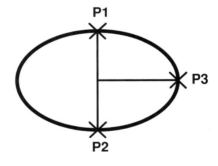

**Specifying <u>Major Axis first</u> (P1/P2),
then Minor Axis (P3)**

**Specifying <u>Minor Axis first</u> (P1/P2),
then Major Axis (P3)**

Continued on the next page...

ELLIPSE....continued

CENTER METHOD

1. Select the **ELLIPSE** command using one of the following:

   Ribbon = Home tab / Draw panel /

   Keyboard = EL <enter> C <enter>

2. The following prompts will appear on the command line:

   Command: _ellipse
   Specify axis endpoint of ellipse or [Arc/Center]: _c
   Specify center of ellipse: *place center of ellipse (P1)*
   Specify endpoint of axis: *place first axis endpoint (either axis) (P2)*
   Specify distance to other axis or [Rotation]: *place the point perpendicular to the first axis (P3)*

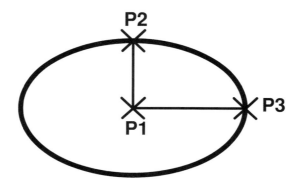

Continued on the next page...

ELLIPSE....continued

ELLIPTICAL ARC METHOD

1. Select the **ELLIPSE** command using one of the following:

    Ribbon = Home tab / Draw panel /

    Keyboard = EL <enter> A <enter>

2. The following prompts will appear on the command line:
    Command: _ellipse
    Specify axis endpoint of ellipse or [Arc/Center]: _a
    Specify axis endpoint of elliptical arc or [center]: *type C <enter>*
    Specify center of axis: *place the center of the elliptical arc (P1)*
    Specify endpoint of axis: *place first axis point (P2)*
    Specify distance to other axis or [Rotation]: *place the endpoint perpendicular to the first axis (P3)*

Specify start angle or [Parameter]: *place the start angle (P4)*

Specify end angle or [Parameter/Included angle]: *place end angle (P5)*

DONUT

A Donut is a circle with *width*. You will define the **Inside** and **Outside** diameters.

Outside Diameter   **Inside Diameter**

1. Select the **DONUT** command using one of the following:

    Ribbon = Home tab / Draw panel /

    Keyboard = DO <enter>

2. The following prompts will appear on the command line:

    Command: _donut
    Specify inside diameter of donut: *type the inside diameter <enter>*
    Specify outside diameter of donut: *type the outside diameter <enter>*
    Specify center of donut or <exit>: *place the center of the first donut*
    Specify center of donut or <exit>: *place the center of the second donut or*
                                             <enter> to stop

Controlling the "FILL MODE"

1. Command: *type FILL <enter>*

2. Enter mode [ON / OFF] <OFF>: *type ON or OFF <enter>*

3. Select type *REGEN <enter>* to regenerate the drawing to show the latest setting
    of the *FILL* mode.

FILL = ON   **FILL = OFF**

POINT

Points are used to locate a point of reference or location. A **Point** may be represented by one of many **Point Styles** shown below in the <u>Point Style box</u>.

The only object snap option that can be used with Point is **Node**.
(Refer to the next page for more information on Node object snap)

HOW TO <u>USE</u> THE POINT COMMAND

1. Select the **POINT** command using one of the following:

   Ribbon = Home tab / Draw panel ▾/ [⬚]

   Keyboard = PO <enter>

2. The following prompts will appear on the command line:

   Command: _point
   Current point modes: PDMODE=3 PDSIZE=0.000
   Specify a point: *place the point location*
   Specify a point: *place another point or press the "ESC" key to stop*

HOW TO <u>SELECT</u> A "POINT STYLE"

1. Open the Point Style dialog box:

   Keyboard = ddptype <enter>

2. The Point Style dialog box will appear.

3. Select a point style tile.

4. Select the **OK** button.

Point Size:
Set Size Relative to Screen
Sets the point display size as a percentage of the screen size. The point display does not change when you zoom in or out

Set Size in Absolute Units
Sets the point display size as the actual units you specify under Point Size. Points are displayed larger or smaller when you zoom in or out.

MORE OBJECT SNAPS

3 MORE OBJECT SNAP OPTIONS:

NODe   This option snaps to the object "**POINT**"
Select the **Node** object snap and place the cursor on the **POINT**.
The cursor will snap to the POINT.
Note: This is the ONLY object snap that you can use with the
object **POINT.**

NEArest   Snaps to the nearest location on an object.
For example, if you want to attach a Line somewhere on a Circle
but not at a quadrant location.
Select the Line command then select the **Nearest** object snap.
Place the cursor anywhere on the circumference of the Circle and
press the left mouse button. The Line will now be accurately
attached to the Circle at the location you selected.

NO TOOL   **MTP**   **Mid Between 2 Points**
Locates a midpoint between two points

You may select this option from the object snap menu (shift+Rt Click) or you may type
MTP <enter> *when prompted for an endpoint.* *No tool* or *running object snap* option is
available.

HOW TO USE "MTP":

1. Draw 2 parallel lines.

2. Select the Line command again.

3. Type **MTP <enter>**

4. Using **Endpoint object snap** select each of the
    2 endpoints (P1) and (P2)

    P1 ————————

    P2 ————————

5. The new line's first endpoint should start <u>exactly</u>
    between those 2 endpoints.

EXERCISE 5A

INSTRUCTIONS:

1. Start a **NEW** file using **2010-Workbook Helper.dwt.**

2. **Draw** the Circle first using the Center / Radius option.

3. Draw the <u>Circumscribed</u> Polygon next using object snap Center and Quadrant object snap to locate the radius.

4. Draw the Lines last using Object snaps Midpoint and Endpoint.

5. **Ortho** (F8) **ON**

6. **Increment Snap** (F9) **OFF** (It will get in your way)

7. Use layer: Object Line

8. **Save** the drawing as: **EX5A**

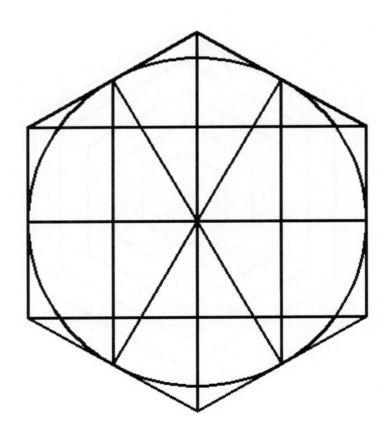

EXERCISE 5B

INSTRUCTIONS:

1. Start a **NEW** file using **2010-Workbook Helper.dwt.**
2. Select the **Point Style** shown below
3. Draw the **Point** first.
4. Draw the <u>Inscribed</u> Polygons
    a. Locate the center of each Polygon using Object Snap **NODE.**
5. **Ortho** (F8) **ON**
6. **Increment Snap** (F9) **OFF**
7. Use layer Object Line
8. **Save** the drawing as: **EX5B**

   Note: Your Point <u>may not</u> appear as large as the one shown below.

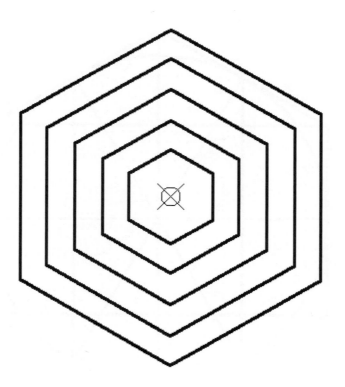

EXERCISE 5C

INSTRUCTIONS:

1. Start a **NEW** file using **2010-Workbook Helper.dwt**.
2. Draw the Objects below using **Ellipse, Line** and **Donut**.
3. Use Object Snap: **Quadrant, Center** and **Nearest**.
4. **Ortho** (F8) **ON**
5. **Increment Snap** (F9) **OFF**
6. Use any layer you like.
7. **Save** the drawing as: **EX5C**

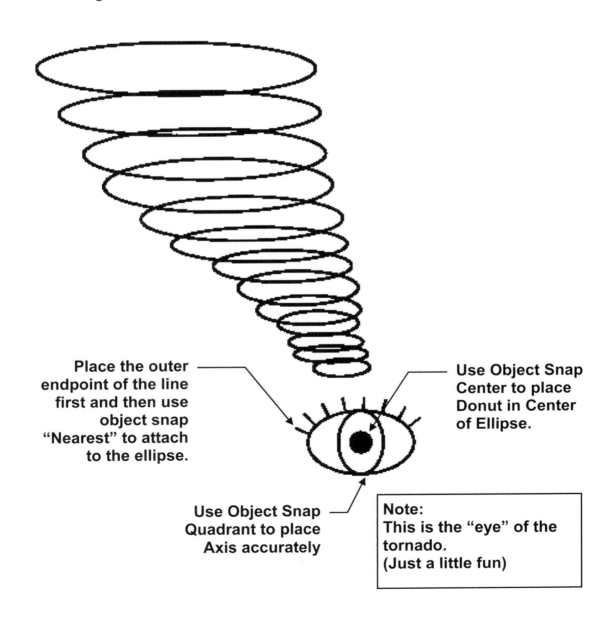

Place the outer endpoint of the line first and then use object snap "Nearest" to attach to the ellipse.

Use Object Snap Center to place Donut in Center of Ellipse.

Use Object Snap Quadrant to place Axis accurately

Note:
This is the "eye" of the tornado.
(Just a little fun)

EXERCISE 5D

INSTRUCTIONS:

1. Start a **NEW** file using **2010-Workbook Helper.dwt.**
2. Select the **Point Style** shown below
3. Draw the **Point** first.
4. Draw the **Ellipses:**
   a. Locate the **center** of each Ellipse using Object Snap: **NODE.**
   B. Locate the **Axis** using Object snap: **Quadrant**
5. **Ortho** (F8) **ON**
6. **Increment Snap** (F9) **OFF**
7. Use layer: Object Line
8. **Save** the drawing as: **EX5D**

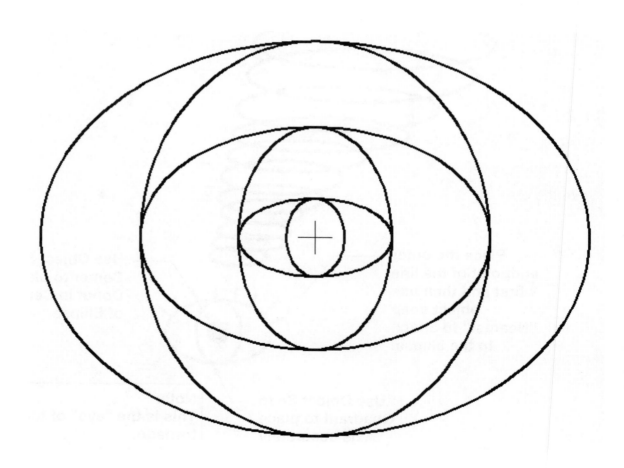

EXERCISE 5E

INSTRUCTIONS:

STEP 1

1. Start a **NEW** file using **2010-Workbook Helper.dwt.**
2. Draw the 3 **DONUTS** shown below.
3. Use Object Snap **Center** to place the centers accurately.
4. **Increment Snap** (F9) **OFF**
5. Use layer: Object Line

Donut sizes

1. ID = 0   OD = 1
2. ID = 1.5   OD = 2
3. ID = 2.5   OD = 3.5

STEP 2

6. Turn the FILL mode OFF
7. Type **Regen <enter>**
8. **Save** the drawing as: **EX5E**

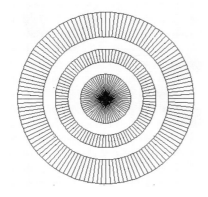

EXERCISE 5F

INSTRUCTIONS:

1. Start a **NEW** file using **2010-Workbook Helper.dwt.**
2. Draw the objects below using: **Point, Polygon, Ellipse** and **Donut.**
3. **Ortho** (F8) **ON**
4. **Increment Snap** (F9) **OFF**
5. Use whatever layers you like
6. **Save** the drawing as: **EX5F**

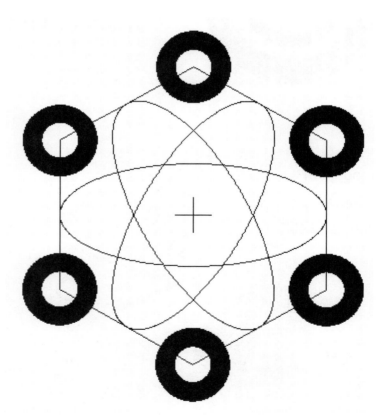

EXERCISE 5G

INSTRUCTIONS:

1. Start a **NEW** file using **2010-Workbook Helper.dwt.**

2. Draw the Zig Zag Lines approximately as shown below.

3. Set <u>Running Object Snap</u> to "**Endpoint**".

4. **Increment Snap** (F9) **OFF**

5. **Ortho** (F8) **= ON**

6. Draw the circles at the "**Midpoint between 2 points**". (P1 and P2)

7. Use layer: Object Line.

8. **Save** the drawing as: **EX5G**

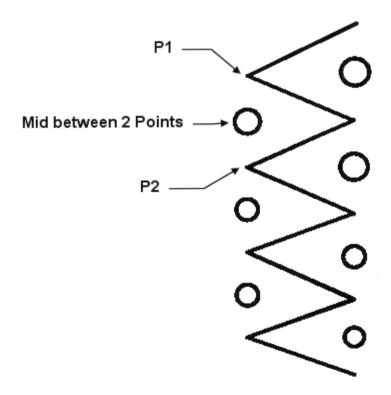

EXERCISE 5H

INSTRUCTIONS:

1. Start a **NEW** file using **2010-Workbook Helper.dwt.**

2. Draw the **Rectangle** on the left first. (any size)

3. Draw the **Polygon** next using the "**EDGE**" option.

    A. Use Object Snap: "**Endpoint**" to place **Edge** points accurately.

4. **Increment Snap** (F9) **OFF**

5. **Ortho** (F8) **= ON**

6. Use Layer: Object Line

7. **Save** the drawing as: **EX5H**

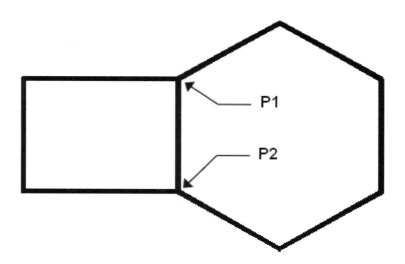

LEARNING OBJECTIVES

After completing this lesson, you will be able to:

1. Use the Break command
2. Trim an object to a cutting edge
3. Extend an object to a boundary
4. Move and object to a new location
5. Explode objects to their primitive entities.

LESSON 6

BREAK

The **BREAK** command allows you to break an object at a single point (Break at Point) or between two points (Break). I think of it as breaking a single line segment into two segments or taking a bite out of an object.

METHOD 1 - Break at a Single Point
How to break one Line into two separate objects with no visible space in between.

1. Select the **BREAK AT POINT** command by using::

   Ribbon = Home tab / Modify panel /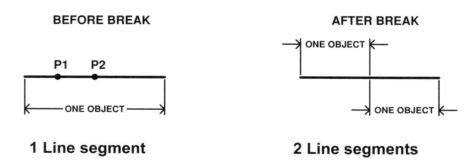

2. _break Select objects: *select the object to break (P1).*

3. Specify first break point: *select break location (P2) accurately.*

BEFORE BREAK	AFTER BREAK

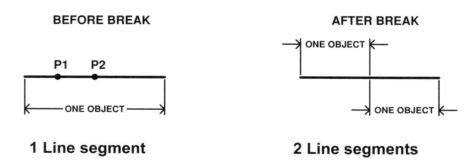

1 Line segment   **2 Line segments**

BREAK....continued

METHOD 2
Break between 2 points. (Take a bite out of an object.)
(Use this method if the <u>location</u> of the BREAK <u>is not important</u>.)

1. Select the **BREAK** command using one of the following:

>   **Ribbon = Home tab / Modify panel /** 🖻

>   **Keyboard = BR <enter>**

2. _break Select object: ***pick the first break location (P1).***
3. Specify second break point or [First point]: ***pick the second break location (P2).***

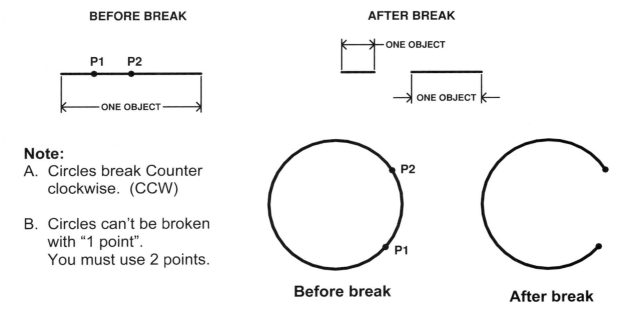

Note:
A. Circles break Counter clockwise. (CCW)

B. Circles can't be broken with "1 point".
    You must use 2 points.

The following method is the same as method 2 above; however, <u>use this method if the location of the break is very specific.</u>

1. Select the **BREAK** command.
2. _break Select objects: ***select the object to break (P1) anywhere on the object.***
3. Specify second break point or [First point]: ***type F <enter>.***
4. Specify first break point: ***select the first break location (P2) accurately.***
5. Specify second break point: ***select the second break location (P3) accurately.***

TRIM

The **TRIM** command is used to trim an object to a **cutting edge**. You first select the "Cutting Edge" and then select the part of the object you want to trim. The object to be trimmed must actually intersect the cutting edge or <u>could</u> intersect if the objects were infinite in length.

1. Select the Trim command using one of the following:

   Ribbon = Home tab / Modify panel /

   Keyboard = TR <enter>

2. The following will appear on the command line:

Command: _trim
Current settings: Projection = UCS   Edge = Extend
Select cutting edges ...
Select objects or <select all>: *select cutting edge(s) by clicking on the object (P1)*
Select objects: *stop selecting cutting edges by pressing the <enter> key*
Select object to trim or shift-select to extend or
[Fence/Crossing/Project/Edge/eRase/Undo]: *select the object that you want to trim. (P*
(Select the part of the object that you want to disappear, not the part you want to remain)
Select object to trim or [Fence/Crossing/Project/Edge/eRase/Undo]: *press <enter> to st*

Object to Trim (P2)

Object to Trim (P2)

Cutting edge (P1)

Before Trim   **After Trim**

Note: You may toggle between Trim and Extend (page 6-5). Hold down the shift key a
the Extend command will activate. Release the shift key and you return to Trim.

Fence See page 6-11
Edge See page 6-5
Project See page 6-5

Crossing You may select objects using a Crossing Window.

eRase You may erase an object instead of trimming while in the Trim command.

Undo You may "undo" the last trimmed object while in the Trim command

EXTEND

The **EXTEND** command is used to extend an object to a **boundary.** The object to be extended must actually or theoretically intersect the boundary.

1. Select the **EXTEND** command using one of the following:

   Ribbon = Home tab / Modify panel / ⊟✎

   Keyboard = EX <enter>

2. The following will appear on the command line:

Command: _extend
Current settings: Projection = UCS   Edge = Extend
Select boundary edges ...
Select objects or <select all>: *select boundary (P1) by clicking on the object.*
Select objects: *stop selecting boundaries by selecting <enter>.*
Select object to extend or shift-select to Trim or
[Fence/Crossing/Project/Edge/Undo]:*select the object that you want to extend (P2
and P3). (Select the end of the object that you want to extend.)*
Select object to extend or [Fence/Crossing/Project/Edge/Undo]:*stop selecting objects
by pressing <enter>.*

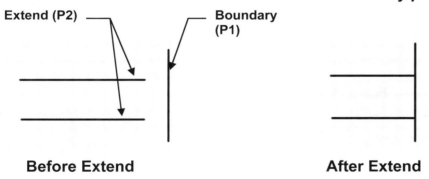

Before Extend   **After Extend**

Note: When selecting the object to be extended (P2 and P3 above) click on the end pointing towards the boundary.

Fence See page 6-11
Crossing See page 6-4
Project   Same as Edge except used only in "3D".

Edge (Extend or No Extend)
   In the **"Extend"** mode, (default mode) the boundary and the Objects to be extended need only intersect if the objects were infinite in length.
   In the **"No Extend"** mode the boundary and the objects to be extended must visibly intersect.

Undo See page 6-4

MOVE

The **MOVE** command is used to move object(s) from their current location (base point) to new location (second displacement point).

1. Select the Move command using one of the following:

      Ribbon = Home tab / Modify panel /

      Keyboard = M <enter>

2. The following will appear on the command line:

Command: _move
Select objects: *select the object(s) you want to move (P1).*
Select objects: *select more objects or stop selecting object(s) by selecting <enter>.*
Specify base point or displacement: *select a location (P2) (usually on the object).*
Specify second point of displacement or <use first point as displacement>: *move the object to its* <u>*new location*</u> *(P3) and press the left mouse button.*

Warning: If you press <enter> instead of actually picking a new location (P3) with the mouse, Autocad will send it into <u>*Outer Space*</u>*. If this happens just select undo tool and try again.*

Base Point (P2)

P1

Drag to new location (P3)

Original Location

New Location

EXPLODE

The **EXPLODE** command changes (explodes) an object into its primitive objects.
For example: A rectangle is originally one object. If you explode it, it changes into 4 lines.
Visually you will not be able to see the change unless you select one of the lines.

1. Select the **Explode** command by using one of the following:

   Ribbon = Home tab / Modify panel /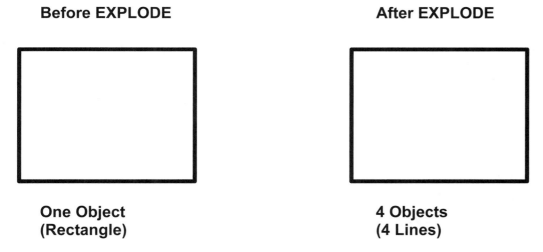

   Keyboard = X <enter>

2. The following will appear on the command line:

Command: _explode
Select objects: *select the object(s) you want to explode.*
Select objects: *select <enter>.*

Before EXPLODE	After EXPLODE
One Object (Rectangle)	4 Objects (4 Lines)

(Notice there is no visible difference. But now you have 4 lines instead of 1 Rectangle)

Try this:
Draw a rectangle and then click on it. The entire object highlights.
Now explode the rectangle, then click on it again. Only the line you clicked on should be highlighted. Each line that forms the rectangular shape is now an individual object.

EXERCISE 6A

INSTRUCTIONS:

1. Start a **NEW** file using **2010-Workbook Helper.dwt.**

2. **Draw** the objects below:

      A. Use object snap **Midpoint** to locate the center for the Circles.

      B. **Ortho** (F8) **ON** when drawing Horizontal and Vertical lines.

      C. Turn **Increment Snap** (F9) **ON**

      D. Use Layer: Object Line

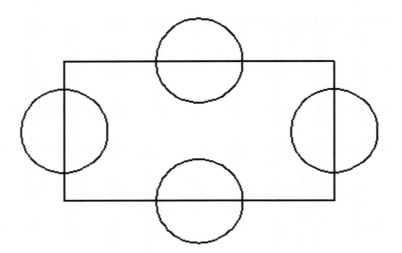

3. Trim the Circles and Rectangle to match the illustration below.

4. **Save** the drawing as: **EX6A**

EXERCISE 6B

INSTRUCTIONS:

1. Start a **NEW** file using **2010-Workbook Helper.dwt.**

2. **Draw** the objects below:

    A. **Circles** should have the same center

    B. **Ortho** (F8) **ON** when drawing Horizontal and Vertical lines.

    C. **Ortho** (F8) **OFF** when drawing lines on an Angle.

    D. Turn **Increment Snap** (F9) **OFF**

    E. Use Layer: Object Line

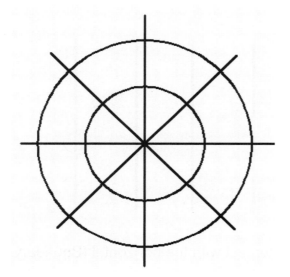

3. **TRIM** the Lines to match the illustration below.

4. **Save** the drawing as: **EX6B**

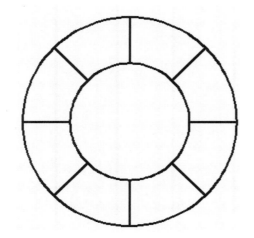

EXERCISE 6C

INSTRUCTIONS:

1. Start a **NEW** file using **2010-Workbook Helper.dwt.**

2. **Draw** the **Lines** below exactly as shown.

    A. **Ortho** (F8) **ON** when drawing Horizontal and Vertical lines.

    B. Turn **Increment Snap** (F9) **ON**

    C. Turn **Osnap** (F3) **OFF**

    D. Use Layer: Object Line

3. **Extend** the vertical lines to intersect with the horizontal (Boundary) Line as shown below.

4. **Save** the drawing as: **EX6C**

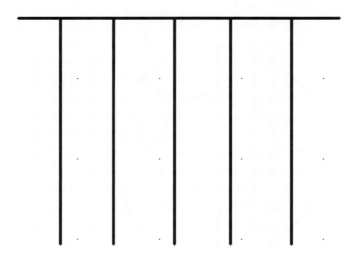

EXERCISE 6D

INSTRUCTIONS:

1. Start a **NEW** file using **2010-Workbook Helper.dwt.**

2. Draw the **LINES** below exactly as shown.

3. Select the "**Extend**" command.

4. Select the "**Boundary**" <enter>

5. Now instead of clicking on each vertical line, type **F <enter>.**

6. Place the cursor approximately at location **P1** and click.

7. Move the cursor to approximately location **P2** and click.

8. Press **<enter>** and **<enter>** again.

9. **Save** the drawing as: **EX6D**

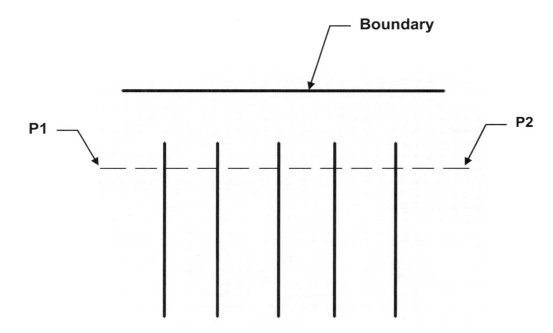

Note:
Be careful to place P1 and P2 above the midpoint of the vertical lines. That instructs AutoCAD to extend the lines in that direction toward the boundary. If you place P1 and P2 below the midpoint of the vertical line, AutoCAD will look for a boundary in that direction. This will confuse AutoCAD because you did not select a boundary below.

EXERCISE 6E

INSTRUCTIONS:

1. Start a **NEW** file using **2010-Workbook Helper.dwt.**
2. **Draw** the objects below using:

   Rectangles, Circles and Lines (for the X's)

   Use Layer: Object Line

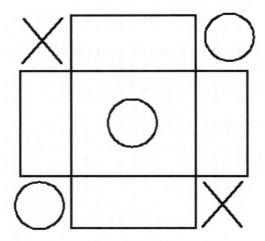

3. Modify the drawing above to appear as the drawing shown below using:

   Explode and **Erase**

   Note: You must Explode a Rectangle before you can Erase an individual line.

4. **Save** the drawing as: **EX6E**

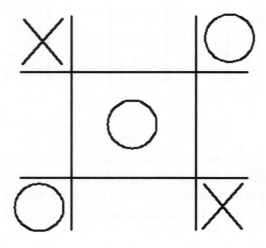

EXERCISE 6F

INSTRUCTIONS:

1. Start a **NEW** file using **2010-Workbook Helper.dwt.**

2. **Draw** the objects below using all of the commands you have learned in the previous lessons.

      Use whatever layers you desire

3. **Save** the drawing as: **EX6F**

Have some fun with this one. But be very accurate.

All objects must intersect exactly. Zoom in and take a look to make sure.

Use 2point Circle and Trim

**Donuts with Fill turned OFF
(Yours may not appear
exactly the same as shown)**

Notes:

LEARNING OBJECTIVES

After completing this lesson, you will be able to:

1. Copy objects
2. Create a mirrored image
3. Add rounded corners
4. Add Chamfered corners

LESSON 7

COPY

The **COPY** command creates a duplicate set of the objects selected.
The COPY command is similar to the MOVE command.

The steps required are:
      1. Select the objects to be copied,
      2. Select a base point
      3. Select a New location for the New copy.

The difference between Copy and Move commands:
The Move command merely moves the objects to a new location.
The Copy command makes a copy and you select the location for the new copy.

1. Select the Copy command using one of the following commands:

    Ribbon = Home tab / Modify panel /

    Keyboard = CO <enter>

2. The following will appear on the command line:

Command: _copy
Select objects: *select the objects you want to copy*
Select objects: *stop selecting objects by selecting <enter>*
Current settings: Copy mode = Multiple
Specify base point or [Displacement/mOde] <Displacement>: *select a base point (P1)*
Specify second point of displacement or <use first point as displacement>: *select the*
new location (P2) for the first copy
Specify second point or [Exit / Undo] <Exit>: *select the new location (P2) for the next*
copy or press <enter> to exit.

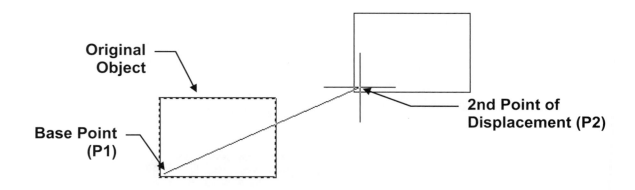

Original Object

Base Point (P1)

2nd Point of Displacement (P2)

Continued on the next page...

COPY....continued

Note:
If you select the option **mOde**, you may select <u>Single</u> or <u>Multiple</u> copy mode.
The default setting is Multiple.
It is practical to leave the mode setting at Multiple. If you choose to make only one copy just press <enter> to exit the Copy command.

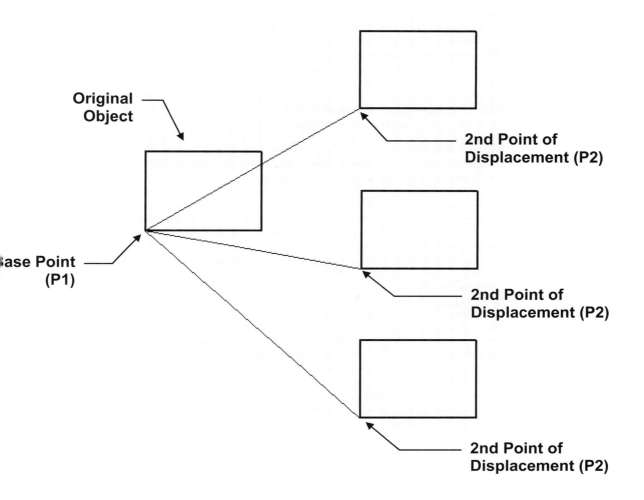

Original Object

2nd Point of Displacement (P2)

Base Point (P1)

2nd Point of Displacement (P2)

2nd Point of Displacement (P2)

The copy command continues to make copies until you press <enter> to exit.

MIRROR

The **MIRROR** command allows you to make a mirrored image of any objects you select. You can use this command for creating right / left hand parts or draw half of a symmetrical object and mirror it to save drawing time.

1. Select the **MIRROR** command using one of the following:

       Ribbon = Home tab / Modify panel / ◿◺

       Keyboard = MI <enter>

2. The following will appear on the command line:

Command: _mirror
Select objects: ***select the objects to be mirrored***
Select objects: ***stop selecting objects by selecting <enter>***
Specify first point of mirror line: ***select the 1st end of the mirror line***
Specify second point of mirror line: ***select the 2nd end of the mirror line***
Erase source objects? [Yes/No] <N>: ***select Y or N***

Mirror Line "Vertical"

Continued on the next page...

MIRROR....continued

Note:
The placement of the "Mirror Line" is important. You may make a mirrored copy
<u>horizontally</u>, <u>vertically</u> or <u>on an angle</u>. See examples below and on the previous page.

**Mirror Line
"Horizontal"**

**Mirror Line
"Angled"**

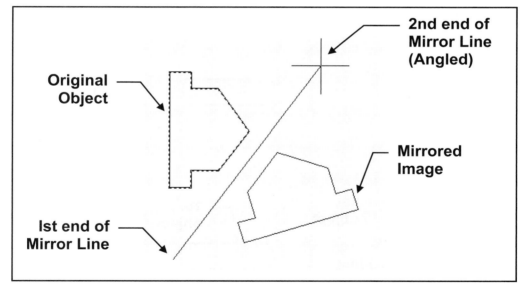

How to control text when using the Mirror command:
(Do the following before you use the mirror command)

1. At the command line type   **mirrtext <enter>**
2. If you <u>want</u> the text to mirror: type 1 <enter>
   If you <u>do not want</u> the text to mirror: type 0 <enter>

Ɉ = ꓷИITTƎƧ TXƎTЯЯIM   MIRRTEXT SETTING = 0

FILLET

The **FILLET** command will create a radius between two objects. The objects do not have to be touching. If two parallel lines are selected, it will construct a full radius.

RADIUS A CORNER

1. Select the **FILLET** command using one of the following:

   Ribbon = Home tab / Modify panel / ⌐

   Keyboard = F <enter>

2. The following will appear on the command line:

3. **Set the radius of the fillet**
Command: _fillet
Current settings: Mode = TRIM, Radius = 0.000
Select first object or [Undo/Polyline/Radius/Trim/Multiple]: *type "R" <enter>*
Specify fillet radius <0.000>: *type the radius <enter>*

4. **Now fillet the objects**
Select first object or [Undo/Polyline/Radius/Trim/Multiple]: *select the 1st object to be filleted*
Select second object or shift-select to apply corner: *select the 2nd object to be filleted*

Before

After

1st object — Fillet — 2nd object

1st object — 2nd object — Fillet

1st object — Fillet — 2nd object

Polyline: This option allows you to fillet all intersections of a Polyline in one operation, such as all 4 corners of a rectangle.

Trim: This option controls whether the original lines are trimmed to the end of the Arc or remain the original length. (Set to Trim or No trim)

Multiple: Repeats the fillet command until you press <enter> or esc key.

FILLET....continued

The FILLET command may also be used to create a square corner.

SQUARE CORNER

1. Select the **FILLET** command using one of the following:

2. The following will appear on the command line:

3. **Select the two lines to form the square corner**
Select first object or [Undo/Polyline/Radius/Trim/Multiple]: *select the 1st object (P1)*

Select second object or shift-select to apply corner: *Hold the shift key down while selecting the 2nd object (P2)*

ORIGINAL

P1

P2

P2

P1

TRIM SHORT ENDS   **TRIM LONG ENDS**

Note: The corner trim direction depends on which end of the object you select.

CHAMFER

The **CHAMFER** command allows you to create a chamfered corner on two lines. There are two methods: **Distance (below) and Angle (next page)**.

DISTANCE METHOD

Distance Method requires input of a distance for each side of the corner.

1. Select the **CHAMFER** command using one of the following:

    Ribbon = Home tab / Modify panel /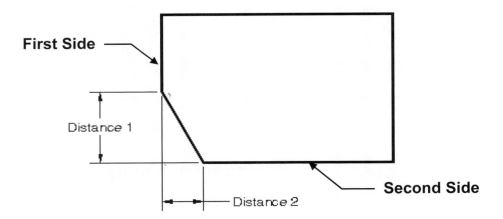

    Keyboard = CHA <enter>

    Command: _chamfer
    (TRIM mode) Current chamfer Dist1 = 0.000, Dist2 = 0.000
    Select first line or [Undo/Polyline/Distance/Angle/Trim/mEthod/Multiple]: **select "D"<enter>.**
    Specify first chamfer distance <0.000>: **type the distance for first side <enter>.**
    Specify second chamfer distance <1.000>: **type the distance for second side <enter>.**

2. **Now chamfer the object.**
    Select first line or [Undo/Polyline/Distance/Angle/Trim/mEthod/Multiple]: **select the (First side) to be chamfered (distance 1).**
    Select second line or shift-select to apply corner: **select the (Second side) to be chamfered (distance 2).**

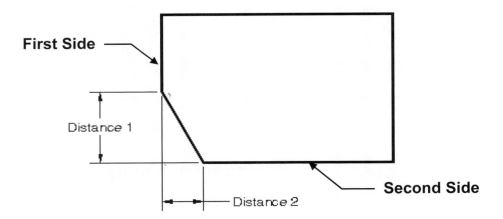

Polyline: This option allows you to Chamfer all intersections of a Polyline in one operation. Such as all 4 corners of a rectangle.

Trim: This option controls whether the original lines are trimmed or remain after the corners are chamfered. (Set to Trim or No trim.)

mEthod: Allows you to switch between **Distance** and **Angle** method. The distance or angle must have been set previously.

Multiple: Repeats the Chamfer command until you press <enter> or esc key.

CHAMFER....continued

ANGLE METHOD

Angle method requires input for the length of the line and an angle

1. Select the **CHAMFER** command

   Command: _chamfer
   (TRIM mode) Current chamfer Dist1 = 1.000, Dist2 = 1.000
   Select first line or [Undo/Polyline/Distance/Angle/Trim/method/Multiple]: *type A <enter>*
   Specify chamfer length on the first line <0.000>: *type the chamfer length <enter>*
   Specify chamfer angle from the first line <0>: *type the angle <enter>*

2. **Now Chamfer the object**
   Select first line or [Undo/Polyline/Distance/Angle/Trim/mEthod/Multiple]: *select the (First Line) to be chamfered. (the length side)*
   Select second line or shift-select to apply corner: *select the (second line) to be chamfered. (the Angle side)*

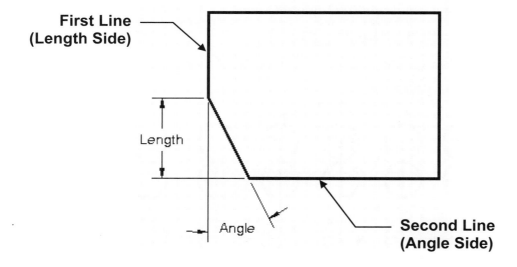

First Line
(Length Side)

Length

Angle

Second Line
(Angle Side)

EXERCISE 7A

INSTRUCTIONS:

1. Start a **NEW** file using **2010-Workbook Helper.dwt.**

2. **Draw** the Lines and one Circle shown below.

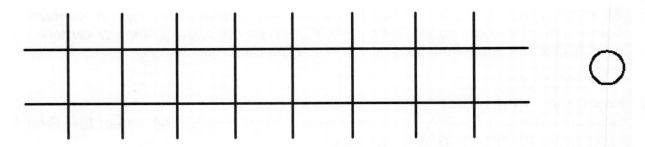

BEFORE COPY

3. Copy the Circle to all of the other locations.

    A. Select the Copy command

    B. Select the Circle

    C. Select the basepoint on the Original Circle.

    Note: the basepoint is different for 3 rows.

    D. Select the New location (2nd point of displacement)

4. **Save** the drawing as: **EX7A**

AFTER COPY

Note: Step by step instructions on the next page.

Helpful hints for 7A

1. Select the Copy command
2. Select the Circle
3. Select the Basepoint: the Center of the Circle (use object snap Center)
4. Set Running Object snap to Endpoint and Intersection and place all 24 accurately.

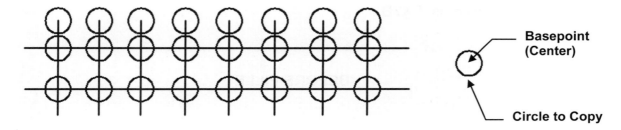

Basepoint (Center)

Circle to Copy

1. Select the Copy command again.
2. Select the Circle to copy
3. Select the Basepoint: the Quadrant (12 o'clock) this time.
4. Place the 8 bottom circles accurately using Endpoint object snap.

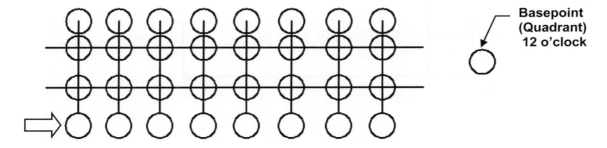

Basepoint (Quadrant) 12 o'clock

1. Select the Copy command
2. Select the Circle to copy
3. Select the Basepoint: the Quadrant (3 o'clock) this time.
4. Place the 2 circles on the left accurately using Endpoint object snap.

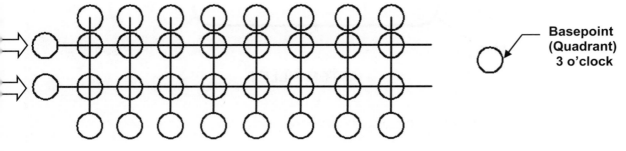

Basepoint (Quadrant) 3 o'clock

You should be able to figure out the last 2 circles on the right.

EXERCISE 7B

INSTRUCTIONS:

1. Start a **NEW** file using **2010-Workbook Helper.dwt.**
2. **Draw** the Rectangle shown below.
3. Radius the corners using the **Fillet** command.
4. **Save** the drawing as: **EX7B**

BEFORE FILLET

AFTER FILLET

EXERCISE 7C

INSTRUCTIONS:

1. Start a **NEW** file using **2010-Workbook Helper.dwt.**
2. **Draw** the 3 Rectangles shown below.
3. Chamfer the corners using the **Chamfer** command.
4. **Save** the drawing as: **EX7C**

BEFORE CHAMFER

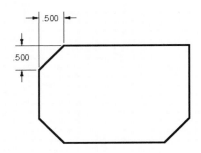

AFTER CHAMFER

EXERCISE 7D

INSTRUCTIONS:

1. Start a **NEW** file using **2010-Workbook Helper.dwt.**
2. **Draw** the Rectangle shown below.
3. Chamfer the corners using the **Chamfer** command.
4. **Save** the drawing as: **EX7D**

BEFORE CHAMFER

AFTER CHAMFER

EXERCISE 7E

INSTRUCTIONS:

1. Start a **NEW** file using **2010-Workbook Helper.dwt.**
2. **Draw** the "Half house shown below.
   Use Layers: Roof, Wall, Window, Door, Plants and Bricks.

Hint: Draw this line first.
Also snap to it for the Mirror Line.
Your halves should match perfectly.

Note: <u>You must be very accurate.</u>
All objects must intersect perfectly.
Use ALL the commands that you have learned.
Consider Trim, Extend, Copy, Rectangle,
2 pt. Circle, Running Object snap: Nearest,
Perpendicular, Quadrant and Endpoint.
Zoom in and out.

3. Now create a Mirrored image of the half house:
   A. Select the Mirror command
   B. Select the objects to be Mirrored
   C. Select the 1st endpoint of the Mirror line (Ortho ON)
   D. Select the 2nd endpoint of the Mirror line.
   E. Answer NO to "Erase Source Objects?"
4. **Save** the drawing as: **EX7E**

Can you imagine how
this might save you
valuable time?

Notes:

LEARNING OBJECTIVES

After completing this lesson, you will be able to:

1. Add a Single Line of text to the drawing
2. Add a paragraph using Multiline text
3. Control tabs, indents and use the Spelling Checker
4. Add Columns
5. Edit existing text
6. Mask the text background

LESSON 8

SINGLE LINE TEXT

SINGLE LINE TEXT allows you to draw one or more lines of text. The text is visible as you type. To place the text in the drawing, you may use the default **START POINT** (the lower left corner of the text), or use one of the many styles of justification described on the next page.

USING THE DEFAULT START POINT

1. Select the **SINGLE LINE TEXT** command using one of the following:

   Ribbon = Annotation tab / Text panel / [A]

   Keyboard = DT <enter>

Command: _dtext
Current text style: "STANDARD" Text height: 0.250 Annotative: No
Specify start point of text or [Justify/Style]: *Place the cursor where the text should start and left click.*
Specify height <0.250>: *type the height of your text <enter>*
Specify rotation angle of text <0>: *type the rotation angle then <enter>*
Enter text: *type the text string; press enter at the end of the sentence*
Enter text: *type the text string; press enter at the end of the sentence*
Enter text: *type the next sentence or press <enter> to stop*

USING JUSTIFICATION

If you need to be very specific, where your text is located, you must use the Justification option. For example if you want your text in the middle of a rectangular box, you would use the justification option "Middle".

The following is an example of Middle justification.

1. Draw a Rectangle 6" wide and 3" high.
2. Draw a Diagonal line from one corner to the diagonal corner.

3. Select the SINGLE LINE TEXT command
   Command: _dtext
   Current text style: "STANDARD" Text height: 0.250
4. Specify start point of text or[Justify/Style]: *type "J"<enter>*
5. Enter an option [Align/Fit/Center/Middle/Right/TL/ TC/TR/ ML/MC/MR/BL/BC/BR]: *type M <enter>*
6. Specify middle point of text: *snap to the midpoint of the diagonal line*
7. Specify height <0.250>: *1 <enter>*
8. Specify rotation angle of text <0>: *0 <enter>*
9. Enter text: *type: HHHH <enter>*
10. Enter text: *press <enter> to stop*

Also refer to Exercise 8C for "Midpoint between 2 pts" method.

SINGLE LINE TEXT....continued

OTHER JUSTIFICATION OPTIONS:

ALIGN
Aligns the line of text between two points specified.
The height is adjusted automatically.

FIT
Fits the text between two points specified.
The height is specified by you and does not change.

CENTER
This is a tricky one. Center is located at the bottom center of Upper Case letters.

MIDDLE
If only uppercase letters are used **MIDDLE** is located in the middle, horizontally and vertically. If both uppercase and lowercase letters are used **MIDDLE** is located in the middle, horizontally and vertically, but considers the lowercase letters as part of the height.

RIGHT
Bottom right of upper case text.

TL, TC, TR
Top left, Top center and Top right of upper and lower case text

ML, MC, MR
Middle left, Middle center and Middle right of upper case text.
(Notice the difference between "Middle" and "MC"

BL, BC, BR
Bottom left, Bottom center and Bottom right of lower case text.
Notice the different location for **BR** and **RIGHT** shown above.
BR considers the lower case letters with tails as part of the height.

MULTILINE TEXT

MULTILINE TEXT command allows you to easily add a sentence, paragraph or tables. The Mtext editor has most of the text editing features of a word processing program. You can underline, bold, italic, add tabs for indenting, change the font, line spacing, and adjust the length and width of the paragraph.

When using MText you must first define a text boundary box. The text boundary box is defined by entering where you wish to start the text (first corner) and approximately where you want to end the text (opposite corner). It is very similar to drawing a rectangle. The paragraph text is considered one object rather than several individual sentences.

USING MULTILINE TEXT

1. Select the MULTILINE TEXT command using one of the following:

    Ribbon = Annotation tab / Text panel /

    Keyboard = MT <enter>

The command line will lists the current style, text height and annotative setting.

   Mtext Current text style: "STANDARD"   Text height: .250 Annotative: No

Note: <u>Annotative</u> will be discussed in lesson 26 and 27.

The cursor will then appear as crosshairs with the letters "abc" attached. These letters indicate how the text will appear using the current font and text height.

2. Specify first corner: ***Place the cursor at the upper left corner of the area where you want to start the new text boundary box and press the left mouse button. (P1)***

3. Specify opposite corner or [Height / Justify / Line Spacing / Rotation / Style / Width / Columns]: ***Move the cursor to the right and down (P2) and press left mouse button.***

MULTILINE TEXT....continued

The Text Formatting Tool bar will appear.

Close

Text Formatting Tool Bar

Text Area Box

Drag to adjust
Height of paragraph

Drag to adjust
width of paragraph

The **Text Formatting toolbar** allows you to select the Text Style, Font, Height etc.
You can add features such as bold, italics, underline and color.

The **Text Area box** allows you to enter the text, add tabs, indent, adjust left hand
margins and change the width and height of the paragraph.

4. After you have entered the text in the Text Area box, select the **Close Text Editor**
   tool.

HOW TO CHANGE THE "abc", ON THE CROSSHAIRS, TO OTHER LETTERS.

You can personalize the letters that appear attached to the crosshairs using the
MTJIGSTRING system variable. (10 characters max) The letters will simulate the
appearance of the font and height selected but will disappear after you place the lower
right corner (P2).

1. Type **MTJIGSTRING** <enter> on the command line.
2. Type the new letters <enter>.

The letters will be saved to the computer, not the drawing. They will appear anytime
you use Mtext and will remain until you change them again.

TABS, INDENTS and SPELLING CHECKER

TABS

Setting and removing Tabs is very easy.

The default setting for tabs is 1". (You may set as many tabs as you need.)
Set or change the stop positions at anytime, using one of the following methods.

Place the cursor on the "Ruler" where you want the tab and left click. A little dark "**L**" will appear. The tab is set.
If you would like to remove a tab, just click and drag it off the ruler and it will disappear

INDENTS

Sliders on the ruler show indention relative to the left side of the text boundary box.
The top slider indents the first line of the paragraph, and the bottom slider indents the other lines of the paragraph.

You may change their positions at anytime, using one of the following methods.

Place the cursor on the "Slider" and click and drag it to the new location.

SPELLING CHECKER

If you have Spell check ON you will be alerted as you enter text with a red line under the misspelled word. Right click on the word and AutoCAD will give you some choices.

You may also check the spelling of your entire drawing using AutoCAD's Spelling Checker.

1. Select **Annotation tab / Text panel ◢** /

   The Check Spelling dialog box will appear.

2. Select Start.
   If AutoCAD finds any words misspelled it will suggest a change.
   You may select **Change** or **Ignore**.
   When finished a message will appear stating "**Spelling Check Complete**".

COLUMNS

STATIC COLUMNS

1. Right click in the **Text Box Area** and select **Columns**

2. Select **Column Settings…**

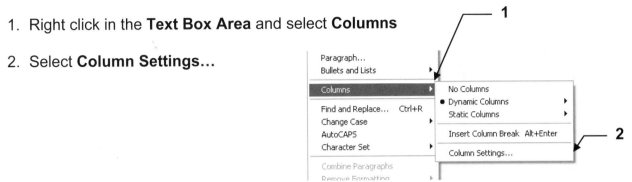

The Column Dialog box appears.

3. Select **Static Columns**

4. Select:
   A. Column Number
   B. Height
   C. Width
   D. Gutter

5. Select the **OK** button

6. The Text Area should appear as shown below with 2 columns divided with a gutter.

7. Start typing in the left hand box. When you fill the left hand box the text will start to spill over into the right hand box.

You may also adjust make changes to width and height using the drag tools.

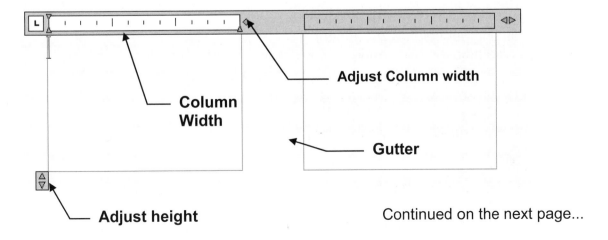

Continued on the next page...

COLUMNS....continued

DYNAMIC COLUMNS

1. Right click in the **Text Box Area** and select **Columns**

2. Select **Column Settings...**

The Column Dialog box appears.

3. Select **Dynamic Columns**

4. Select:
   A. Height
   B. Width
   C. Gutter

5. Select the **OK** button

6. The Text Area will first appear with one column with the width and height you set.

7. When the text fills the first column another column will appear. When the second

column fills another column will appear.
You may also adjust make changes to width and height using the drag tools.

PARAGRAPH and LINE SPACING

PARAGRAPH and LINE SPACING

You may set the tabs, indent and line spacing for individual paragraphs.

1. Right click in the **Text Box Area** and select **Paragraph**.

  The Paragraph dialog box will appear.

EDITING TEXT

SINGLE LINE TEXT

Editing **Single Line Text** is somewhat limited compared to Multiline Text. In the example below you will learn how to edit the text within a Single Line Text sentence. (In Lesson 12 you will learn additional options for editing Single Line text by using the Properties command.)

1. Double click on the <u>Single Line text</u> you want to edit. The text will highlight.

2. Make the changes in place then press \<enter\> \<enter\>.

MULTILINE TEXT

Multiline Text is as easy to edit as it is to input originally. You may use any of the text options shown on the Text Editor tab.

1. Double click on the Multiline text you want to edit.

2. Highlight the text, that you want to change, using click and drag.

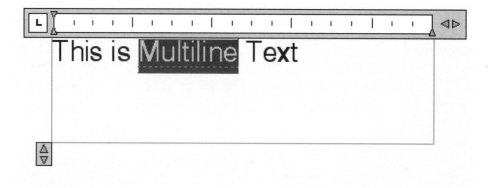

3. Make the changes then select the **Close Text Editor** tool.

EDITING TEXT….continued

You may edit many other Multiline Text features.

Right click in the **Text Box Area.**

The menu shown on the right will appear.

BACKGROUND MASK

Background mask inserts an opaque background so that objects under the text are covered. (masked) The mask will be a rectangular shape and the size will be controlled by the "Border offset factor". The color can be the same as the drawing background or you may select a different color.

1. After selecting the "Background Mask" option from the menu shown on the right, the dialog box shown below will appear.

2. Turn this option **ON** by selecting the "**Use background Mask**" box.

3. Enter a value for the "Border offset factor". The value is a factor of the text height. 1.0 will be exactly the same size as the text. 1.5 (the default) extends the background by 1.5 times the text height. The width will be the same width that you defined for the entire paragraph.

4. In the Fill Color area, select the "Use drawing background color" box if you want to make the background the same color as the drawing background. To specify another color, uncheck this box and select a color. (Note: if you use a color you may need to adjust the "draworder" using Tools / Draworder to bring the text to the front.)

EXERCISE 8A

INSTRUCTIONS:

1. Start a **NEW** file using **2010-Workbook Helper.dwt.**
2. Duplicate the text shown below using **Single Line Text**
3. Use Layer Text.
4. Follow the instructions in each block of text. To start the text in the correct location that is stated in each example move the cursor while watching the coordinate display.
5. **Save** the drawing as: **EX8A**

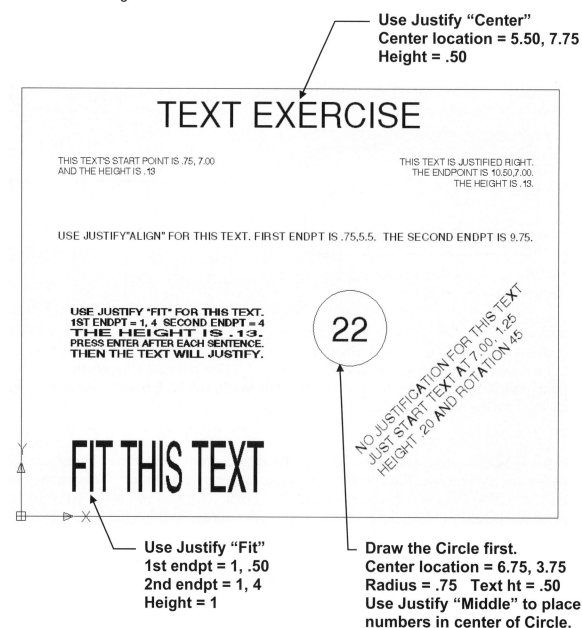

Use Justify "Center"
Center location = 5.50, 7.75
Height = .50

TEXT EXERCISE

THIS TEXT'S START POINT IS .75, 7.00
AND THE HEIGHT IS .13

THIS TEXT IS JUSTIFIED RIGHT.
THE ENDPOINT IS 10.50,7.00.
THE HEIGHT IS .13.

USE JUSTIFY"ALIGN" FOR THIS TEXT. FIRST ENDPT IS .75,5.5. THE SECOND ENDPT IS 9.75.

USE JUSTIFY "FIT" FOR THIS TEXT.
1ST ENDPT = 1, 4 SECOND ENDPT = 4
THE HEIGHT IS .13.
PRESS ENTER AFTER EACH SENTENCE.
THEN THE TEXT WILL JUSTIFY.

22

NO JUSTIFICATION FOR THIS TEXT
JUST START TEXT AT 7.00, 1.25
HEIGHT .20 AND ROTATION 45

FIT THIS TEXT

Use Justify "Fit"
1st endpt = 1, .50
2nd endpt = 1, 4
Height = 1

Draw the Circle first.
Center location = 6.75, 3.75
Radius = .75   Text ht = .50
Use Justify "Middle" to place
numbers in center of Circle.

EXERCISE 8B

INSTRUCTIONS:

1. Start a **NEW** file using **2010-Workbook Helper.dwt.**

2 Duplicate the text shown below using <u>**MULTILINE**</u> text.

3. Use Layer **Text.**

4. Select **MULTILINE text**

5. Use Text Style: **Standard**

6. Use font: **SansSerif**

7. Text Height: **.250**

8. Follow the directions below. You may make change the settings as you type or you
   may enter all of the text and then go back and edited it. Your choice.

9. Enter all text shown below.

10. Save as: **EX8B**

The following is an exercise for tabs, indent, bold and underline.
>   1. This sentence should be indented 1 inch.
>   a. This sentence should be indented 1.50 inch.

And now back to the left margin.

Isn't this fun?

STUDENT NAME	STUDENT ID	GRADE
Susie Que	1234567	A
John Smith	8910116	B

**Set tabs to 4.50 and 8.00
Clear all previous tabs.**

EXERCISE 8C

INSTRUCTIONS:
1. Start a **NEW** file using **2010-Workbook Helper.dwt.**

The following exercise is designed to teach you how to insert text into the exact middle of a rectangular area. You will use both <u>Single Line Text</u> and <u>Multiline Text</u>.

2. Follow the instructions for each text box below.
3. Save as: **EX8C**

SINGLE LINE TEXT
1. Draw a 6" wide by 3" high rectangle.
2. Select "**SINGLELINE TEXT**"
3. Use Justify: Middle
4. Use "**M2P**" object snap to locate the middle of the rectangle. (Refer to 5-8)
    a. Type **m2p** <enter> on the command line.
    b. Using object snap "Endpoint" snap to **(P1)** corner and then the diagonal corner **(P2)**
5. Use Text Ht: 1"
6. Rotation "0"
7. Type the word "Middle" and <enter><enter

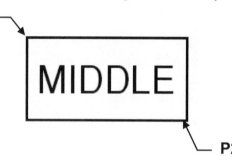

MULTILINE TEXT
1. Draw another 6" wide by 3" high rectangle.
2. Select "**MULTILINE TEXT**"
3. Start the Text boundary box at the upper left corner **(P1)** of the rectangle. Place the opposite corner at the lower right corner **(P2)** of the rectangle. (Use "Endpoint" object snap to be accurate.)
4. Select **Standard, 1.00, Sans Serif and Middle Center.**
5. Type **MIDDLE** and select **Close Text Editor.**

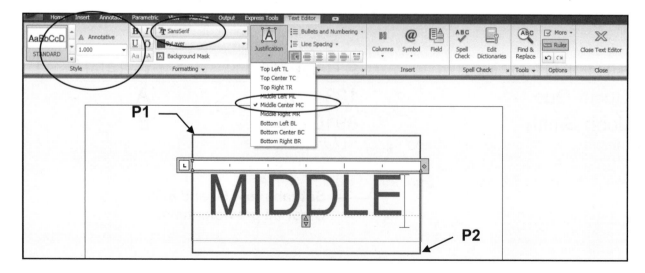

EXERCISE 8D

INSTRUCTIONS:

1. Start a **NEW** file using **2010-Workbook Helper.dwt.**

2. Draw **two** 6" long lines as shown.

3. Select Single Line Text
   a. Select **Justify - Center**.
   b. Use Midpoint snap to place the justification point at the midpoint of the line.
   c. Use text height 1" and rotation angle 0.
   d. Type the word "**Happy**" <enter> <enter> **(Use upper and lower case)**

4. Select Single Line Text again.
   a. This time select **Justify - BC**. (bottom center)
   b. Use Midpoint snap to place the justification point at the midpoint of the line.
   c. Use text height 1" and rotation angle 0.
   d. Type the word "Happy" <enter> <enter> (Use upper and lower case)

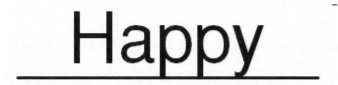

Notice the difference between <u>Center</u> and <u>Bottom Center</u>?
"Center" only considers the Upper Case letters when justifying.
"Bottom Center" is concerned about those Lower Case letters.
Can you see how you could accidentally place your text too high or too low? Think about the difference between Center and Bottom Center.

5. Save as: **EX8D**

EXERCISE 8E

INSTRUCTIONS:

1. Start a **NEW** file using **2010-Workbook Helper.dwt.**

2. Draw the 2 sentences below using **Single Line text**. (Use Layer **Text**)

> Mirror Text setting 1

> Mirror Text setting 0

3. Change the **Mirrtext** setting:

   A. Type **mirrtext <enter>**

   B. Type **1 <enter>**

4. Using the Mirror command, mirror the **first sentence** using a vertical mirror line

   (P1 and P2)

5. Now change the **Mirrtext** command setting to **0**

6. Mirror the **second** sentence.

Notice the difference in the mirrored copy. Sometimes you will want the mirrored text to be reversed and sometimes you will not. Now you know how to control it.

7. Save as **EX8E**

LEARNING OBJECTIVES

After completing this lesson, you will be able to:

1. Understand the Origin
2. Draw objects accurately using Coordinate Input
3. Input Absolute and Relative coordinates
4. Input using Direct Distance Entry
5. List information about an object
6. Measure the distance between two points
7. Identify a location within the drawing
8. Create your own 8-1/2 x 11 Master Border
9. Print from Model Space

LESSON 9

COORDINATE INPUT

In the previous lessons you have been using the cursor to place objects.
In this lesson you will learn how to place objects in <u>specific locations</u> by entering coordinates. This process is called **Coordinate Input**.

This is not difficult, so do not start to worry.

Autocad uses the ***Cartesian Coordinate System.***
The Cartesian Coordinate System has 3 axes, X, Y and Z.

The **X** is the Horizontal axis. *(Right and Left)*
The **Y** is the Vertical axis. *(Up and Down)*
The **Z** is Perpendicular to the X and Y plane.
*(The **Z** axis, <u>which is not discussed in this workbook</u>, is used for 3D.)*

UCS Icon

Look at the User Coordinate System (UCS) icon in the lower left corner of your screen.
The X and Y arrows are pointing in the positive direction.

The location where the X , Y and Z axes intersect is called the **ORIGIN**.
*The **Origin** always has a coordinate value of X=0, Y=0 and Z=0 (0,0,0)*
When you move the cursor away from the Origin, in the direction of the arrows, the X and Y coordinates are positive.
When you move the cursor in the opposite direction, the X and Y coordinates are negative.

Using this system, every point on the screen can be specified using positive or negative X and Y coordinates.

ABSOLUTE COORDINATES

There are 3 types of Coordinate input, **Absolute, Relative** and **Polar**.

(Relative is discussed on the next page and Polar will be discussed in Lesson 11)

ABSOLUTE COORDINATES

When inputting absolute coordinates the input format is: **X, Y**   (that is: X comma Y)

Absolute coordinates come *from the ORIGIN* and are typed as follows: **8, 5**

The first number (8) represents the **X-axis** (horizontal) distance from the Origin and the second number (5) represents the **Y-axis** (vertical) distance from the Origin.
The two numbers must be separated by a **comma**.

An absolute coordinate of **4, 2** will be **4** units to the right (horizontal) and **2** units up (vertical) from the current location of the Origin.

An absolute coordinate of **-4, -2** will be **4** units to the left (horizontal) and **2** units down (vertical) from the current location of the Origin.

The following are examples of Absolute Coordinate input.
Notice where the Origin is located in each example.

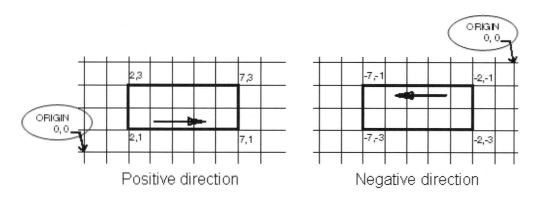

Positive direction   Negative direction

Very important:

While working on Lessons 9 and 10 it is best to **turn off** Dynamic Input.

DYN, on the status bar, should be gray.

Refer to Lesson 11 for more information on Dynamic Input.

RELATIVE COORDINATES

RELATIVE COORDINATES

Relative coordinates come *from the last point entered*. (Not from the Origin)

The first number represents the **X-axis** (horizontal) and the second number represents the **Y-axis** (vertical) just like the absolute coordinates.

To distinguish the relative coordinates from absolute coordinates the two numbers must be <u>preceded by an @ symbol</u> in addition to being separated by a **comma**.

A Relative coordinate of **@5, 2** will go to the **right** 5 units and **up** 2 units <u>from the last point entered</u>.

A Relative coordinate of **@-5, -2** will go to the **left** 5 units and **down** 2 units <u>from the last point entered</u>.

The following is an example of Relative Coordinate input.

Start here with Absolute
Coordinates 2,1

Very important:

While working on Lessons 9 and 10 it is best to **<u>turn off</u>** Dynamic Input.

DYN, on the status bar, should be gray.

Refer to Lesson 11 for more information on Dynamic Input.

EXAMPLES OF COORDINATE INPUT

Scenario 1.
If you want to draw a line with the first endpoint "at the Origin" and the second endpoint
3 units in the positive X direction.

0,0 —————— 3,0

1. Select the Line command.
2. You are prompted for the first endpoint: **Type 0, 0 <enter>**
3. You are then prompted for the second endpoint: **Type 3, 0 <enter>**

What did you do?
The first endpoint coordinate input, 0,0 means that you do not want to move away from
the Origin. You want to start "**ON**" the Origin.

The second endpoint coordinate input, 3, 0 means that you want to move 3 units in the
positive X axis. The "0" means you do not want to move in the Y axis. So the line will be
exactly horizontal.

Scenario 2.
You want to start a line 8 units directly above the origin and it will be 4 units in length,
perfectly vertical.

@0,4

1. Select the Line command.
2. You are prompted for the first endpoint: **Type 0, 8 <enter>**
3. You are prompted for the second endpoint: **Type @0, 4 <enter>**

0,8

What did you do?
The first endpoint coordinate input, 0, 8 means you do not want to move in the X axis
direction but you do want to move in the Y axis direction.

The second endpoint coordinate input @0, 4 means you do not want to move in the X
axis "from the last point entered" but you do want to move in the Y axis "from the last
point entered. (Remember the @ symbol is only necessary if you are entering
coordinates on the command line.)

Scenario 3.
Now try drawing 5 connecting line segments.
1. Select the Line command.
2. First endpoint: 2, 4 <enter>
3. Second endpoint: @ 2, -3 <enter>
4. Second endpoint: @ 0, -1 <enter>
5. Second endpoint: @ -1, 0 <enter>
6. Second endpoint: @ -2, 2 <enter>
7. Second endpoint: @ 0, 2 <enter> <enter>

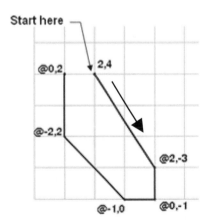

Note: If you enter an incorrect coordinate, just select the undo arrow and the last
segment will disappear and you will have another chance at entering the correct
coordinate.

DIRECT DISTANCE ENTRY (DDE)

DIRECT DISTANCE ENTRY is a combination of keyboard entry and cursor movement. **DDE** is used to specify distances in the horizontal or vertical axes <u>from the last point entered</u>. **DDE** is a ***Relative Input.*** Since it is used for Horizontal and Vertical movements, **Ortho** <u>must be</u> **ON**.

(Note: to specify distances on an angle, refer to Polar Input in Lesson 11)

**Using DDE is simple. Just move the cursor and type the distance.
Negative and positive is understood automatically by moving the cursor up (positive), down (negative), right (positive) or left (negative) <u>from the last point entered.</u> No minus sign necessary.**

Moving the cursor to the right and typing 5 <enter> tells AutoCAD that the 5 is positive and Horizontal.
Moving the cursor to the left and typing 5 <enter> tells AutoCAD that the 5 is negative and Horizontal.
Moving the cursor up and typing 5 <enter> tells AutoCAD that the 5 is positive and Vertical.
Moving the cursor down and typing 5 <enter> tells AutoCAD that the 5 is negative and Vertical.

EXAMPLE:

1. <u>Ortho must be ON</u>.
2. Select the Line command.
3. Type: 1, 2 <enter> to enter the first endpoint using Absolute coordinates.
4. Now move your cursor to the right and type: 5 <enter>
5. Now move your cursor up and type: 4 <enter>
6. Now move your cursor to the left and type: 5 <enter> <enter> to stop

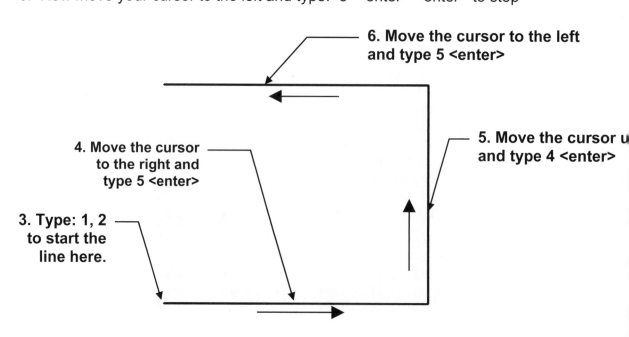

6. Move the cursor to the left and type 5 <enter>

5. Move the cursor u and type 4 <enter>

4. Move the cursor to the right and type 5 <enter>

3. Type: 1, 2 to start the line here.

MEASURE TOOLS and ID Point

The following tools are very useful to confirm the location or size of objects.

The Measure tools enables you to measure the distance, radius, angle area, or volume of a selected object. The default option is Distance.

1..You may access these tools as follows:

Ribbon = Home tab / Utilities Panel / Distance ▼

2. Select one of the tools and follow the instructions on the command line.

ID Point
The ID Point command will list the X and Y coordinates of the point that you select. The coordinates listed will be from the Origin.

1. Select the ID Point command by typing: **ID <enter>**
2. Select a location point, such as the endpoint of a line.
    The X and Y coordinates of the endpoint will be displayed.

Example:
1. Command: **id<enter>**
2. Snap to the endpoint
3. Coordinates, from the Origin, are displayed.

2. Snap to endpoint

Specify point: X = 6.8293   Y = 6.6926   Z = 0.0000

EXERCISE 9A

INSTRUCTIONS:

1. Start a **NEW** file using
   2010-Workbook Helper.dwt

2. Set **Units**
   Units = Decimal
   Precision = 0.000

3. Set **Drawing Limits**
   Lower Left Corner = 0,0
   Upper Rt. Corner = 11, 8.5

4. Show the new limits using:
   Zoom / All

5. **Important**: Change Lineweight settings
   to **Inches** and **adjust Display scale**.
   Refer to bottom of page 3-13

6. Set **Grids** and **Snap**
   Snap = .25
   Grids = 1.00

7. Draw the border below using the
   dimensions shown.
   Use **Layer = Borderline**

8. Place the Text as shown (See pg. 8-2)
   Use **Layer = Text**
   Text Ht. = .250

9. Save this border as **Border A**

10. **Plot**
    Refer to "Basic Plotting from Model
    Space" on the next page.

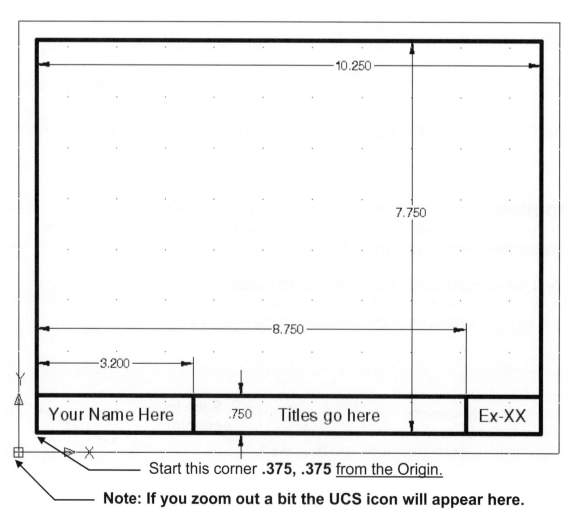

Your Name Here   .750   Titles go here   Ex-XX

Start this corner **.375, .375** from the Origin.

Note: If you zoom out a bit the UCS icon will appear here.

Basic Plotting from Model Space

Note: More Advance plotting methods will be explained in Lessons 26 and 27.

1. **Important:** Open the drawing you want to plot.
2. Select: **Zoom / All** to center the drawing within the plot area.
3. Select the Plot command using one of the following;

Quick Access toolbar =

Ribbon = Output tab / Plot Panel /

Keyboard = Plot <enter>

The Plot –Model dialog box will appear.

4. More Options button.

4. Select the **"More Options"** button to expand the dialog box.

The dialog box expands
To show more options.

Continued on the next page...

5. Select your printer from the drop down list or "Default windows system printer.pc3"
   Note: If your printer is not shown in the list you should configure your printer within AutoCAD. This is not difficult. Refer to Appendix-A for step by step instructions.

6. Select the Paper size :**Letter**.

7. Select the Plot Area **Extents**

8. Select the Plot Offset **Center the Plot**

9. Select Plot Scale **1 : 1**

10. Select the Plot Style table **Monochrome.ctb** (for all black)
    Acad.ctb (for color)

11. If the following box appears, select **Yes**

12. Select the **Preview** button.

Continued on the next page...

Your Name Here	Titles go here	Ex-XX

Does your display appear as shown above?
If yes, press <enter> and proceed to 13.
If not, recheck 1 through 11.

You have just created a **Page Setup**. All of the settings you have selected can now be saved. You will be able to recall these settings for future plots using this page setup. To save the **Page Setup** you need to **ADD** it to the model tab within this drawing.

13

13. Select the **ADD** button.

14. Type the new Page Setup name **Class Model A**

14

15. Select the **OK** button.   **15**

Continued on the next page...

New Page Setup name appears

16. Select the **Apply to Layout** button.

16 ⟋   ⟋ 17

17. Select the **OK** button to send the drawing to the printer or select Cancel if you do not want to print the drawing at this time. The Page Setup will still be saved.

18. Save the entire drawing again using **Save as: Border A**.
    The Page Setup will be saved to the drawing and available to select in the future. You will not have to select all the individual settings again unless you choose to change them.

Now go tape your print out to the refrigerator door.
It is quite an accomplishment.

EXERCISE 9B

INSTRUCTIONS:

1. Open **Border A**

2. Draw the objects below using Absolute and Relative coordinates.

3. Use Layer = Object line.

4. Edit the Title and Ex-XX by double clicking on the text. Do not erase and replace.

5. Do not dimension (We will get to that soon)

6. Save as **EX9B**

7. Plot using Page Setup **Class Model A** (Refer to page 9-11 and 9-12)

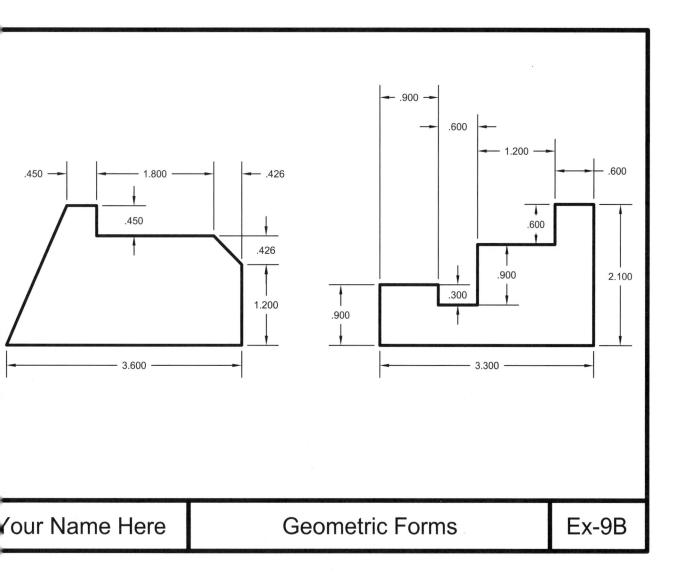

| Your Name Here | Geometric Forms | Ex-9B |

EXERCISE 9C

INSTRUCTIONS:

1. Open **Border A**

2. Draw the objects below using Direct Distance Entry (DDE).

3. Use **Layer = Object line**.

4. Edit the Title and Ex-XX by double clicking on the text. Do not erase and replace.

5. Do not dimension

6. Save as **EX9C**

7. Plot using Page Setup **Class Model A** (Refer to page 9-11 and 9-12)

| Your Name Here | DDE Practice | Ex-9C |

LEARNING OBJECTIVES

After completing this lesson, you will be able to:

1. Move the Origin
2. Control the display of the UCS icon

LESSON 10

MOVING THE ORIGIN

As previously stated in Lesson 9, the **ORIGIN** is where the X, Y, and Z-axes intersect. The Origin's (0,0,0) default location is in the lower left-hand corner of the drawing area. But you can move the Origin anywhere on the screen using the UCS command. (The default location is designated as the "**World**" option or WCS. When it is moved it is UCS, User Coordinate System.)

You may move the Origin many times while creating a drawing. This is not difficult and will make it much easier to draw objects in specific locations.

Refer to the examples on the next page.

To MOVE the Origin:

1. Select one of following:

   Ribbon = View tab / Coordinates panel /

   Keyboard = ucs <enter>

   Current ucs name: *WORLD*
2. Specify origin of UCS or [Face/NAmed/OBject/Previous/View/World/X/Y/Z/ZAxis] <World>: _o

3. Specify new origin <0,0,0>: *type coordinates or use the cursor to place new Origin location.*

To RETURN the Origin to the default "World" location (the lower left corner):

1. Select one of the following:

   Ribbon = View tab / Coordinates panel /

   Keyboard = ucs <enter> w <enter>

MOVING THE ORIGIN....continued

The following are a few examples of moving the Origin.

R.771

1.448

2.262

If you move the Origin to the lower left corner of the rectangle it will make if very easy to accurately place the center of the circle.

How to place the Circle accurately:
1. Select the "Move Origin" as shown on the previous page
2. Snap to the lower left corner of the rectangle using object snap Endpoint.
   The UCS icon should now be displayed as shown above.
3. Select the Circle command
4. Enter the coordinates to the center of the circle (2.262, 1.448 <enter>)
5. Enter the radius (.771 <enter)

6. Select UCS World to return the UCS icon to the lower left corner.
   (Refer to previous page)

More examples of moving the Origin

DISPLAYING THE UCS ICON

The UCS icon is merely a drawing aid. You can control how it is displayed.
It can be visible (on) or invisible (off). It can move with the Origin or stay in the default location. You can even change its appearance.

1. Select the **View tab / Coordinates panel**

2. Select the **Show UCS Icon at Origin ▼**

The UCS origin icon follows the Origin when you move it's location.
I find it very helpful to know where the Origin is at all times by merely looking for the UCS icon.

The UCS icon is visible.

The UCS icon is invisible.

3. Select **UCS Icon Properties**

When you select this option the dialog box shown below will appear.
You may change the Style, Size and Color of the icon at any time.
Changing the appearance is personal preference and will not affect the drawing or commands.

4. Select the **OK** button.

EXERCISE 10A

INSTRUCTIONS:

1. Open **Border A**

2. Move the Origin to the locations noted before you draw the objects.

3. Use Layer = Object line.

4. Edit the Title and Ex-XX by double clicking on the text. Do not erase and replace.

5. Do not dimension

6. Save as **EX10A**

7. Plot using Page Setup **Class Model A** (Refer to page 9-11 and 9-12)

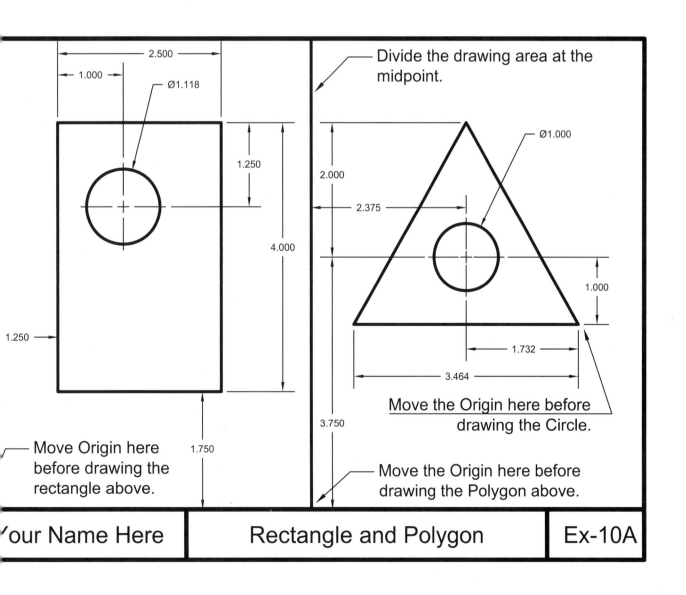

| 'our Name Here | Rectangle and Polygon | Ex-10A |

EXERCISE 10B

INSTRUCTIONS:

1. Open **Border A**

2. Move the Origin to the locations noted before you draw the objects.

3. Use Layer = Object line.

4. Edit the Title and Ex-XX by double clicking on the text. Do not erase and replace.

5. Do not dimension

6. Save as **EX10B**

7. Plot using Page Setup **Class Model A** (Refer to page 9-11 and 9-12)

Move the Origin to th Center befo drawing the small circle

.750

.750

4.000

3.500

2.013

2.000

Ø1.800

4X Ø1.000

1.785

3.256

5.750

2.263

Move the Origin here before drawing the large Circle.

Move the Origin here before drawing the Rectangle.

| Your Name Here | Window Hinge Plate | Ex-10 |

EXERCISE 10C

INSTRUCTIONS:

1. Open **Border A**

2. You decide where to move the origin.

3. Use Layers = Object line, Hidden line and Center line.

4. Edit the Title and Ex-XX by double clicking on the text. Do not erase and replace.

5. Do not dimension

6. Save as **EX10C**

7. Plot using Page Setup **Class Model A** (Refer to page 9-11 and 12)

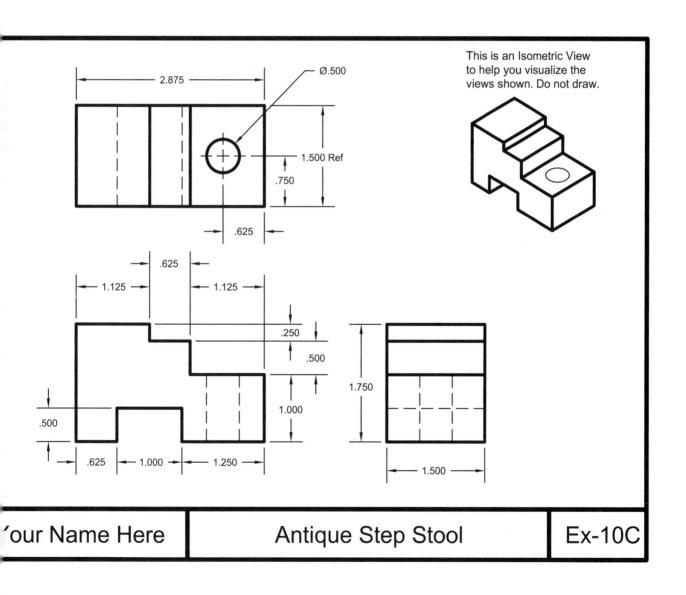

This is an Isometric View to help you visualize the views shown. Do not draw.

| 'our Name Here | Antique Step Stool | Ex-10C |

Notes:

LEARNING OBJECTIVES

After completing this lesson, you will be able to:

1. Understand the Polar Degree Clock
2. Draw Lines to a specific length and angle
3. Draw objects using Polar Coordinate Input
4. Use Dynamic Input
5. Use Polar Tracking and Polar Snap

LESSON 11

POLAR COORDINATE INPUT

In Lesson 9 you learned to control the length and direction of horizontal and vertical lines using Relative Input and Direct Distance Entry. Now you will learn how to control the length and **ANGLE** of a line using <u>**POLAR**</u> Coordinate Input..

UNDERSTANDING THE *"POLAR DEGREE CLOCK"*

Previously when drawing Horizontal and Vertical lines you controlled the direction using a <u>Positive</u> or <u>Negative</u> input. ***Polar Input is different***. The Angle of the line will determine the direction.

For example: If I want to draw a line at a 45 degree angle towards the upper right corner, you would use the angle 45. But if I want to draw a line at a 45 degree angle towards the lower left corner, you would use the angle 225.

You may also use Polar Input for Horizontal and Vertical lines using the angles 0, 90, 180 and 270. No negative input is required.

POLAR DEGREE CLOCK

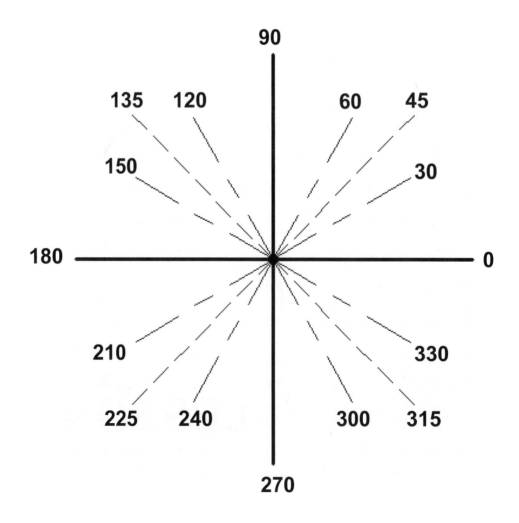

POLAR COORDINATE INPUT....continued

DRAWING WITH *POLAR COORDINATE INPUT*

When entering polar coordinates the first number represents the **Distance** and the second number represents the **Angle**. The two numbers are separated by the **less than (<)** symbol. The input format is: **distance < angle**

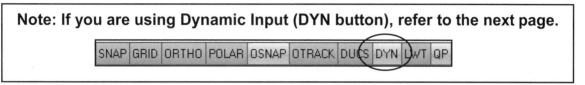

Note: If you are using Dynamic Input (DYN button), refer to the next page.

SNAP | GRID | ORTHO | POLAR | OSNAP | OTRACK | DUCS | DYN | LWT | QP

A Polar coordinate of **@6<45** will be a distance of <u>6 units</u> and an angle of <u>45 degrees</u> *from the last point entered.*

Here is an example of Polar input for 4 line segments.

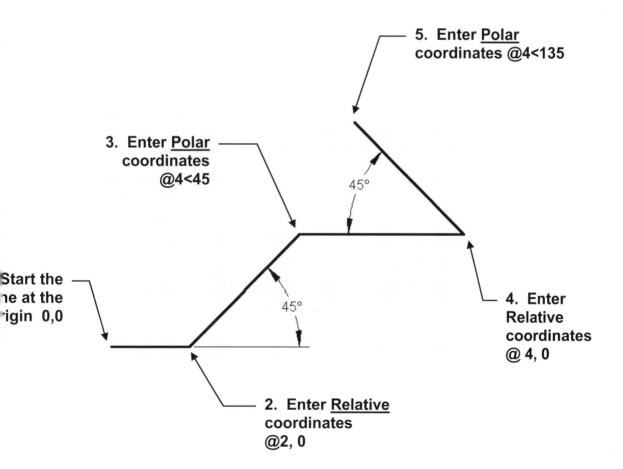

5. Enter <u>Polar</u> coordinates @4<135

3. Enter <u>Polar</u> coordinates @4<45

45°

Start the line at the origin 0,0

45°

4. Enter Relative coordinates @ 4, 0

2. Enter <u>Relative</u> coordinates @2, 0

DYNAMIC INPUT

To help you keep your focus in the "drawing area", AutoCAD has provided a command interface called **Dynamic Input**. You may input information within the Dynamic Input tool tip box instead of on the command line.

When AutoCAD prompts you for the **First point** the Dynamic Input tool tip displays the (Absolute: X, Y) distance from the Origin.
Enter the **X** dimension, <u>press the **Tab key**</u>, enter the **Y** dimension then **<enter>.**

First point

When AutoCAD prompts you for the **Second and Next points** the Dynamic Input tool tip displays the (Relative: Distance and Angle) from the last point entered.
Enter the **distance**, <u>press the **tab key**</u>, move the cursor in the approximate desired angle and enter the **angle** then **<enter>.**

Second and all next points

How to turn Dynamic Input ON or OFF

Select the **DYN** button on the status bar or use the F12 key.

Here is an example of Dynamic input for 4 line segments.

5. Enter <u>Polar</u> coordinates **4 tab 135**

3. Enter <u>Polar</u> coordinates **4 tab 45**

45°

1. Start the line at the Origin 0,0

45°

4. Enter Relative coordinates 4, 0

45°

2. Enter <u>Relative</u> coordinates 2, 0

Note: Refer to Appendix C for more ways to control the Dynamic Input Tool tip box display.

DYNAMIC INPUT....continued

To enter Cartesian coordinates (X and Y)

1. Enter an "**X**" coordinate value <u>and a</u> **comma**.

2. Enter an "**Y**" coordinate value <enter>.

To enter Polar coordinates (from the last point entered)

1. Enter the **distance** value from the last
   point entered.

2. Press the **Tab** key.

3. Move the cursor in the approximate
   direction and enter the **angle** value <enter>

Note: Move the cursor in the approximate
direction and enter an angle value of <u>0-180</u> **only**
Dynamic Input does not required 360 degrees.
(Refer to example on the next page)

How to specify Absolute or Relative coordinates while using Dynamic Input.

To enter **absolute** coordinates when relative coordinate format is displayed in the
tooltip. Enter **#** to temporarily override the setting.

To enter **relative** coordinates when absolute coordinate format is displayed in the
tooltip. Enter@ to temporarily override the setting.

Note about Ortho
You may toggle Ortho **On** and **OFF** by holding down the **shift** key.
This is an easy method to use Direct Distance Entry while using Dynamic Input.

USING DYNAMIC INPUT and POLAR COORDINATES

The following is a simple drawing to practice Dynamic Input and Polar coordinates.
Think about how this differs from the basic polar input on page 11-2 and 11-3.

Example on the next page.

1 Set the Status Bar as follows:
DYN = ON   All others = OFF

2. Select the **Line** command:

3. Start the Line near the lower left corner of
the drawing area.

Line A   4. Move the cursor to the right.
5. Type 2 <tab> 0 <enter>

Line B   6. Move the cursor up and to the right
7. Type 3 <tab> 45 <enter>

Line C   8. Move the cursor up.
9. Type 2 <tab>   90<enter>

Line D   10. Move the cursor down and to the left.
11. Type 4 <tab>   135<enter>   (Note: 180 – 45 = **135**)

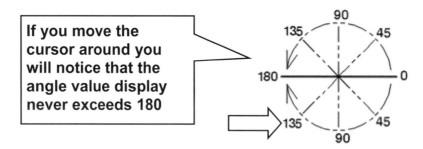

If you move the cursor around you will notice that the angle value display never exceeds 180

Line E   12. Move the cursor to the left.
13. Type 1.293 <tab> 180 <enter>

Line F   14. Move the cursor down.
15. Type 1.293 <tab> 90 <enter>
16. <enter> to stop

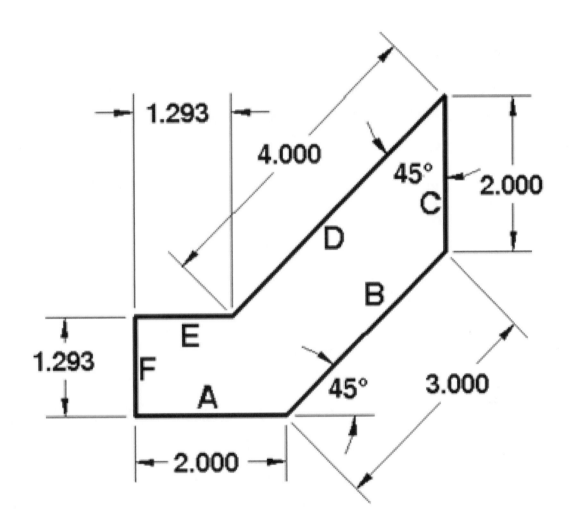

POLAR TRACKING

Polar Tracking can be used instead of **Dynamic Input**. When *Polar Tracking* is "**ON**", a dotted *"tracking"* line and a *"tool tip"* box appear. The tracking line…. "snaps" to a **preset angle increment** when the cursor approaches one of the preset angles.   The word *"Polar"*, followed by the *"distance"* and *"angle"* from the last point appears in the box.   (A step by step example is described on the next page.)

Tracking Line

Polar: 1.219 < 30°

Polar Tool Tip box

HOW TO SET THE INCREMENT ANGLE

1. Right Click on the **POLAR** button on the Status Bar and select "**SETTINGS**", <u>or select an angle from the list.</u>

SNAP | GRID | ORTHO | POLAR | OSNAP | OTRACK | DUCS | DYN | LWT | QP

POLAR ANGLE SETTINGS

Increment Angle   Choose from a list of Angle increments including 90, 45, 30, 22.5, 18, 15,10 and 5. It will also snap to the selected angles multiples. For example: if you choose 30 it will snap to 30, 60, 90, 120, 150, 180, 210, 240, 270, 300, 330 and 0.

Additional Angles   Check this box if you would like to use an angle other than one in the Incremental Angle list. For example: 12.5.

New   You may add an angle by selecting the "New" button. You will be able to snap to this new angle in addition to the incremental Angle selected. But you will not be able to snap to it's multiple. For example, if you selected 7, you would not be able to snap to 14.

Delete   Deletes an Additional Angle. Select the Additional angle to be deleted and then the Delete button.

POLAR ANGLE MEASUREMENT

<u>ABSOLUTE</u> Polar tracking <u>angles</u> are relative to the UCS.

<u>RELATIVE TO LAST SEGMENT</u> Polar tracking <u>angles</u> are relative to the last segment.

SING POLAR TRACKING and DDE

1. Set the Polar Tracking Increment Angle to 15.

2. Turn all the Status Bar buttons to Off except POLAR.
   (Note: DYN must be OFF)

   | SNAP | GRID | ORTHO | POLAR | OSNAP | OTRACK | DUCS | DYN | LWT | QP |

3. Select the Line command.

4. Start the Line anywhere in the drawing area.

5. Move the cursor in the direction of P2 until the Tool Tip box displays 30 degrees.
6. Type 2 <enter> (for the length).

7. Move the cursor in the direction of P3 until the Tool Tip box displays 90 degrees.
8. Type 2 <enter> (for the length).

9. Move the cursor in the direction of P4 until the Tool Tip box displays 0 degrees.
10 Type 2 <enter> (for the length).

11 Move the cursor in the direction of P5 until the Tool Tip box displays 150 degrees
12 Type 2 <enter> (for the length).

13 Move the cursor in the direction of P6 until the Tool Tip box displays 180 degrees.
14 Type 2 <enter> (for the length).

15 Then type C for close.

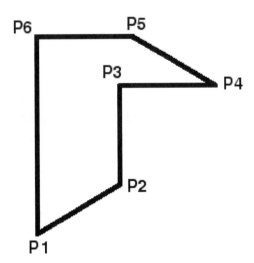

OLAR TRACKING ON or OFF

You may toggle Polar Tracking On or Off using one of the following:
Left click on the POLAR button on the Status Bar or Press F10

POLAR SNAP

Polar Snap is used with Polar Tracking to make the cursor snap to specific *distances* and *angles*. If you set Polar Snap distance to 1 and Polar Tracking to angle 30 you can draw lines 1, 2, 3 or 4 units long at an angle of 30, 60, 90 etc. without typing anything. You just move the cursor and watch the tool tips.

(A step-by-step example is described on the next page)

SETTING THE POLAR SNAP

1. Set the **Polar Tracking Increment Angle** as shown on page 11-8

2. Right Click on the **SNAP** button on the Status Bar and select "**SETTINGS**"

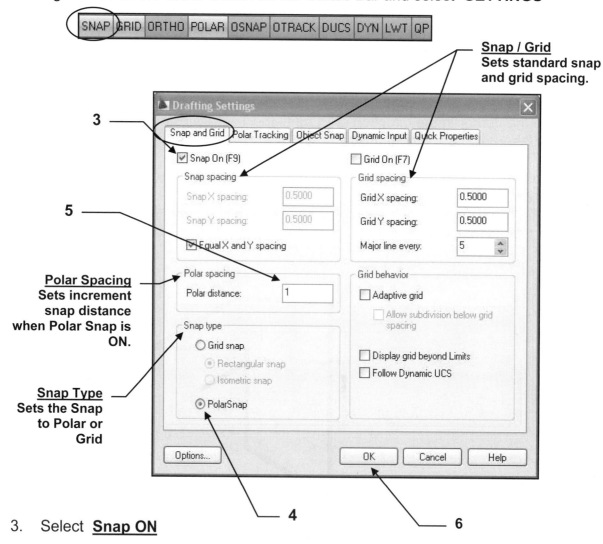

3. Select **Snap ON**

4. Select **PolarSnap**

5. Set the **Polar Distance**

6. Select **OK** button.

USING POLAR TRACKING and POLAR SNAP

Now let's draw the objects below again, but this time with "Polar Snap" instead of Direct Distance Entry.

1. Set **Polar Tracking** Increment Angle to 30 and **Polar Snap** to 1.00.

2. Turn all the Status Bar buttons Off except **SNAP**, **GRID** and **POLAR**.

3. Select the Line command:

P1   4. Start the Line anywhere in the drawing area.

P2   5. Move the cursor in the direction of P2 until the Tool Tip box displays
Polar 2.00 <30°

P3   6. Move the cursor in the direction of P3 until the Tool Tip box displays
Polar 2.00 <90°

P4   7. Move the cursor in the direction of P4 until the Tool Tip box displays
Polar 2.00 <0°

P5   8. Move the cursor in the direction of P5 until the Tool Tip box displays
Polar 2.00 <150°

P6   9. Move the cursor in the direction of P6 until the Tool Tip box displays
Polar 2.00 <180°

10 Then type C for close.

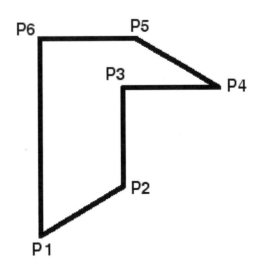

NOTE: You may OVERRIDE the Polar Settings at any time by typing:
Polar coordinates (@ Length< Angle) on the Command line.

EXERCISE 11A

INSTRUCTIONS:

1. Open **Border A**
2. After reviewing the lengths and angles below set the
   Polar Tracking Increment Angle and Polar Snap distance.

   Note: You may have to "override" a few of the lengths. (See bottom of 11-11)

3. Use Layer = Object line.
4. Edit the Title and Ex-XX by double clicking on the text. Do not erase and replace.
5. Do not dimension
6. Save as **EX11A**
7. Plot using Page Setup **Class Model A**

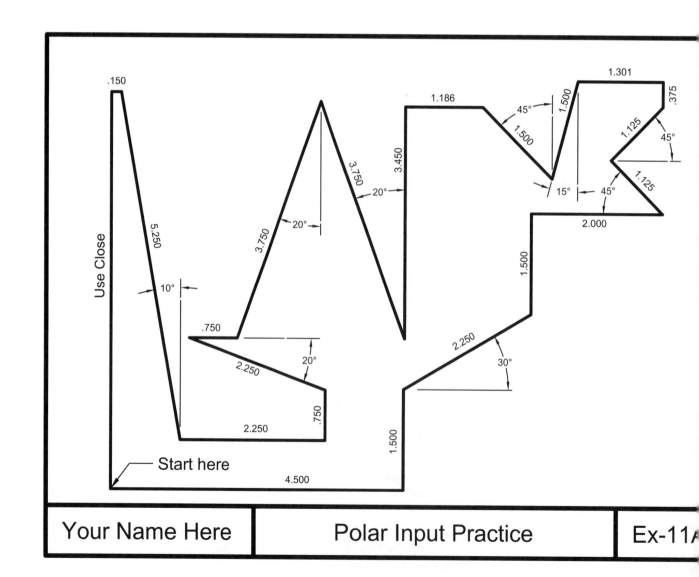

| Your Name Here | Polar Input Practice | Ex-11A |

EXERCISE 11B

INSTRUCTIONS:

1. Open **Border A**
2. After reviewing the lengths and angles below set the
   Polar Tracking Increment Angle to **30**
   Polar Snap distance to **.125**

   Note: the isometric lines are 30, 90, 150, 210, 270 and 330. (Refer to page 11-2)

3. Use Layer = Object line.
4. Edit the Title and Ex-XX by double clicking on the text. Do not erase and replace.
5. Do not dimension
6. Save as **EX11B**
7. Plot using Page Setup **Class Model A**

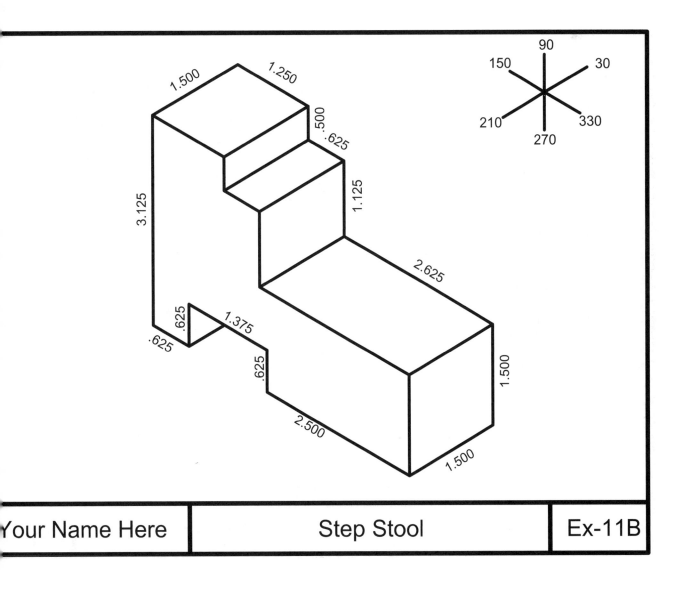

Your Name Here	Step Stool	Ex-11B

EXERCISE 11C

INSTRUCTIONS:

1. Open **Border A**
2. After reviewing the lengths and angles below set the Polar Tracking Increment Angle and Polar Snap distance

   Note: the isometric lines are 30, 90, 150, 210, 270 and 330. (Refer to page 11-2)

3. Use Layer = Object line.
4. Edit the Title and Ex-XX by double clicking on the text. Do not erase and replace.
5. Do not dimension
6. Save as **EX11C**
7. Plot using Page Setup **Class Model A**

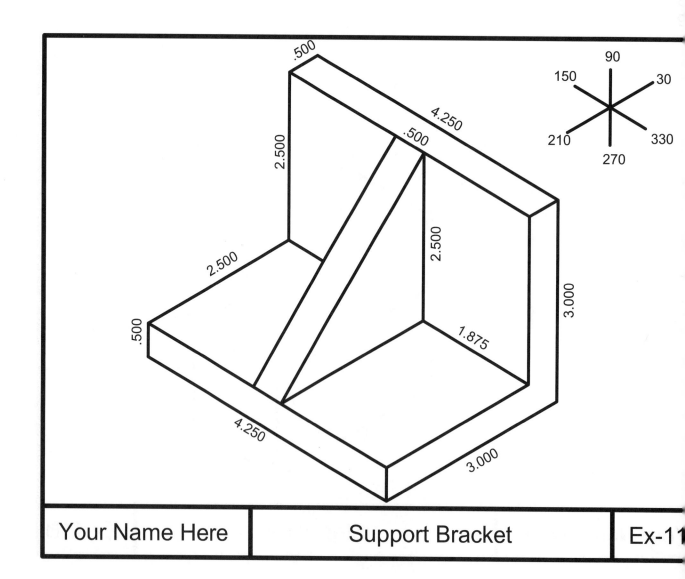

| Your Name Here | Support Bracket | Ex-11 |

EXERCISE 11D

INSTRUCTIONS:

1. Open **Border A**

   Note: You will have to do some adding and subtracting on this one.

2. Draw the 5.000 Diameter Circle using Layer Object line.
3. Draw the 4.000 Diameter Circle using Layer Center line.
4. Draw each polar line on layer Centerline using the following example:

   For example: the line marked "X" is drawn as follows:
   a. Place 1st endpoint in the center of the circle.
   b. Enter Polar coordinates for 2nd endpoint (distance 2.750 angle 138)
   (Refer to Polar Clock on page 11-2)

5. Draw the .500 Diameter Circles. Locate their center using object snap Intersection.
6. Edit the Title and Ex-XX by double clicking on the text. Do not erase and replace.
7. Do not dimension
8. Save as **EX11D**
9. Plot using Page Setup **Class Model A**

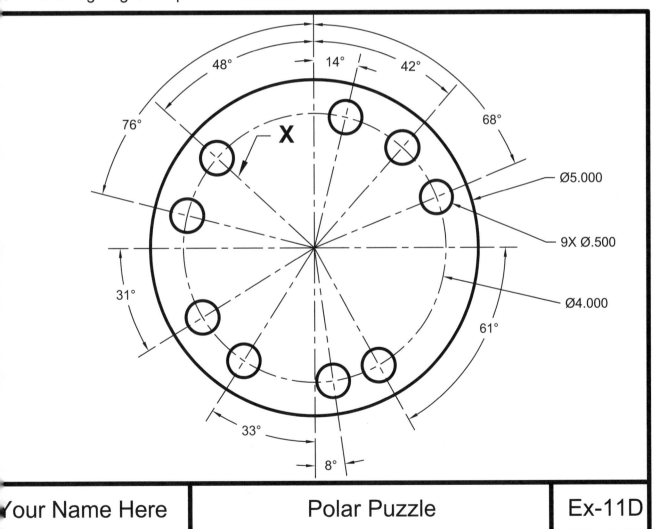

| Your Name Here | Polar Puzzle | Ex-11D |

Notes:

After completing this lesson, you will be able to:

1. Duplicate an object at a specified distance away
2. Make changes to an objects properties
3. Use the Quick Properties Panel

LESSON 12

OFFSET

The **OFFSET** command duplicates an object parallel to the original object at a specified distance. You can offset Lines, Arcs, Circles, Ellipses, 2D Polylines and Splines. You may duplicate the original object or assign the offset copy to another layer.

Examples of Offset objects.

HOW TO USE THE OFFSET COMMAND:

METHOD 1
(Duplicate the Original Object)

1. Select the **OFFSET** command using one of the following:

    Ribbon = Home tab / Modify panel /

    Keyboard = Offset <enter>

2. Command: _offset
   Current settings: Erase source=No Layer=Source OFFSETGAPTYPE=0
   Specify offset distance or [Through/Erase/Layer] <Through>: *type the offset distance or select Erase or Layer. (see <u>options</u> on the next page)*

3. Select object to offset or <Exit/Undo>: *select the object to offset.*

4. Specify point on side to offset or [Exit/Multiple/Undo]<Exit>: *Select which side of the original you want the duplicate to appear by placing your cursor and clicking. (See <u>options</u> on the next page)*

5. Select object to offset or [Exit/Undo]<Exit>: *Press <enter> to stop.*

OFFSET....continued

METHOD 2
Duplicate original object but assign the Offset <u>copy</u> to a different layer)

To automatically place the <u>offset copy</u> on a different layer than the original you must first change the "current" layer to the layer you want the offset copy to be placed on.

1. Select the layer that you want the offset copy placed on from the list of layers.

2. Select the OFFSET command (refer to previous page)

3. Command: _offset
   Current settings: Erase source=No Layer=Source OFFSETGAPTYPE=0
   Specify offset distance or [Through/Erase/Layer] <Through>: *type L <enter>*

4. Enter layer option for offset objects [Current/Source] <Source>: *select C <enter>*

5. Specify offset distance or [Through/Erase/Layer] <Through>: *type the offset dist*
   <enter>

6. Select object to offset or [Exit/Undo] <Exit>: *select the object (source) to offset.*

7. Specify point on side to offset or [Exit/Multiple/Undo]<Exit>: *Select which side of the object you want the duplicate to appear by placing your cursor and clicking. (See <u>options</u> below)*

8. Select object to offset or [Exit/Undo]<Exit>: *Press <enter> to stop.*

OPTIONS:
Through: Creates an object passing through a specified point.

Erase: Erases the source object after it is offset.

Layer: Determines whether offset objects are created on the <u>current</u> layer or on the layer of the <u>source</u> object. Select <u>Layer</u> and then select <u>current</u> or <u>source.</u> (Source is the default)

Multiple: Turns on the multiple offset mode, which allows you to continue creating duplicates of the original without re-selecting the original.

Exit: Exits the Offset command.

Undo: Removes the previous offset copy.

PROPERTIES PALETTE

The **Properties Palette**, shown below, makes it possible to change an object's properties. You simply open the Properties Palette, select an object and you can change any of the properties that are listed.

How to open the Properties Palette:
    Ribbon = Home tab / Properties panel / ↘

    Keyboard = Ctrl + 1

(An example of how to use the Properties Palette is on the next page.)

Close Palette
To close the Palette click on the "X" or press Ctrl + 1

Auto-Hide feature
(Click On and Off here)
The Palette will collapse into a vertical bar if ON. It will reactivate if you pass the cursor over the gray bar.

Multiple selected objects list
If you have selected multiple objects, select individually from this drop down list.

Move
To move, click and drag the gray title bar to a new location.

Resize
To resize, click and drag an edge.

PROPERTIES PALETTE….continued

Example of editing an object using the Properties Palette

1. Draw a 2.00 Radius circle.

2. Open the Properties Palette and select the Circle. (*The Properties for the Circle should appear. You may change any of the properties listed in the Properties Palette for this object. When you press <enter> the circle will change.*)

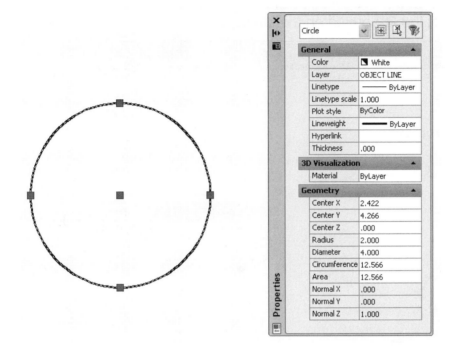

3. Highlight and <u>change the "Radius" to 1.00</u> and the "<u>Layer</u>" to <u>HIDDEN LINE</u> <enter>. *The Circle got smaller and the Layer changed as shown below.*

QUICK PROPERTIES PANEL

The Quick Properties Panel, shown below, will only appear if you have it set to ON. The Quick Properties Panel displays fewer properties and appears when you click once on an object. You may make changes to the objects properties using Quick Properties just as you would using the Properties Palette. (AutoCAD is just giving you another option)

How to turn Quick Properties Panel On or OFF.
Select the **QP** button on the status bar.

Select an object

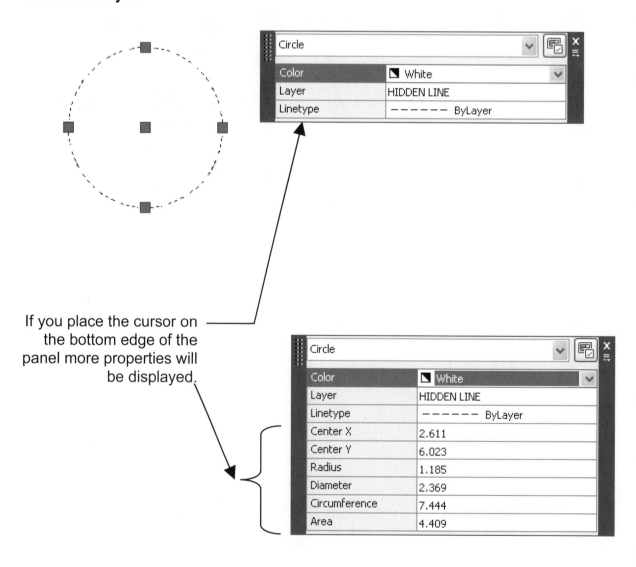

If you place the cursor on the bottom edge of the panel more properties will be displayed.

Refer to the next page to "Customize" the Quick Properties Panel.

CUSTOMIZING THE QUICK PROPERTIES PANEL

You may add or remove properties from the Quick Properties Panel. <u>And it is easy</u>.

1. Select the **Customize** button.

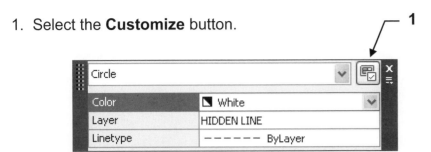

2. Select the Object, from the list, that you would like to customize.

3. <u>Check</u> the boxes for the properties that you <u>want to appear</u>.
   <u>Uncheck</u> the boxes that you <u>do not want to appear</u>.

4. Select **OK**

Note: The customizing is saved to the computer not the drawing file.

OFFSETGAPTYPE

When you offset a closed 2D object, such as a rectangle, to create a **larger** object it results in potential gaps between the segments. The **offsetgaptype** system variable controls how these gaps are closed.

To set the offsetgaptype:

Type: ***offsetgaptype <enter>***

Enter one of the following:

0 = Fills the gap by extending the polyline segments. (default setting)

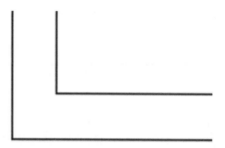

1 = Fills the gap with <u>filleted arc segments</u>. The radius of each arc segment is equal to the offset distance.

2 = Fills the gap with <u>chamfered line segments.</u> The perpendicular distance to each chamfer is equal to the offset distance.

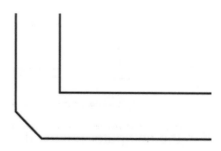

EXERCISE 12A

INSTRUCTIONS:

1. Open **Border A**
2. Draw the Objects below using Circles, Rectangles and Lines.
3. Use the Offset command to offset the objects.
4. Use Layers = Object line and Hidden line.
5. Edit the Title and Ex-XX by double clicking on the text. Do not erase and replace.
6. Do not dimension
7. Save as **EX12A**
8. Plot using Page Setup **Class Model A**

Offset distance .250

Ø4.000
Ø2.500
Ø.700

4.000
.500 Offset
2.500

The solid lines are on layer Object line and 1.00 in length.
(Remember Polar Snap from Lesson 11?)
The dashed lines are on Layer Hiddenline
The Offset is .250. Use Trim and Extend after Offsetting.

Your Name Here | Offset Exercise | Ex-12A

EXERCISE 12B

INSTRUCTIONS:

1. Open **Border A**
2. Follow the steps shown on the next page to draw the Window Schedule shown below.
3. Use Layers = Object line and Text.
4. Edit the Title and Ex-XX by double clicking on the text. Do not erase and replace.
5. Do not dimension
6. Save as **EX12B**
7. Plot using Page Setup **Class Model A**

WINDOW SCHEDULE

SYM	SIZE	TYPE
(A)	5'-0" X 4'-0"	WOOD FIXED
(B)	3'-0" X 4'-0"	WOOD FIXED
(C)	3'-0" X 3'-0"	WOOD FIXED
(D)	2'-0" X 3'-0"	ALUMINUM SLIDER

Your Name Here	Window Schedule	Ex-12

EXERCISE 12B
HELPER

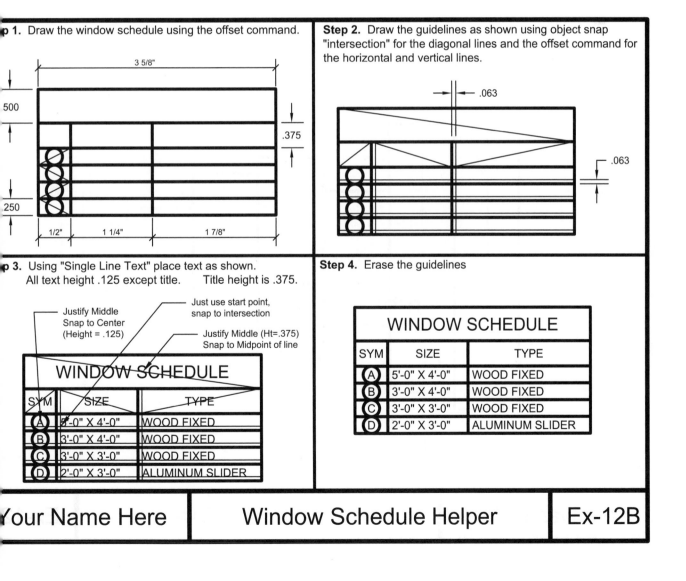

Step 1. Draw the window schedule using the offset command.

3 5/8"

500

.375

250

1/2"   1 1/4"   1 7/8"

Step 2. Draw the guidelines as shown using object snap "intersection" for the diagonal lines and the offset command for the horizontal and vertical lines.

.063

.063

Step 3. Using "Single Line Text" place text as shown. All text height .125 except title.   Title height is .375.

Justify Middle
Snap to Center
(Height = .125)

Just use start point,
snap to intersection

Justify Middle (Ht=.375)
Snap to Midpoint of line

WINDOW SCHEDULE

SYM	SIZE	TYPE
A	5'-0" X 4'-0"	WOOD FIXED
B	3'-0" X 4'-0"	WOOD FIXED
C	3'-0" X 3'-0"	WOOD FIXED
D	2'-0" X 3'-0"	ALUMINUM SLIDER

Step 4. Erase the guidelines

WINDOW SCHEDULE

SYM	SIZE	TYPE
A	5'-0" X 4'-0"	WOOD FIXED
B	3'-0" X 4'-0"	WOOD FIXED
C	3'-0" X 3'-0"	WOOD FIXED
D	2'-0" X 3'-0"	ALUMINUM SLIDER

Your Name Here	Window Schedule Helper	Ex-12B

EXERCISE 12C

INSTRUCTIONS:
1. Open **Border A**
2. Draw a horizontal and vertical line on Layer Object line.

3. Draw the upper half using the Offset command.

4. MIRROR the upper half. (Do not include the horizontal centerline)
5. Change the horizontal line to layer Center line using Properties Palette or QP.
6. Edit the Title and Ex-XX by double clicking on the text. Do not erase and replace.
7. Do not dimension
8. Save as **EX12C**
9. Plot using Page Setup **Class Model A**

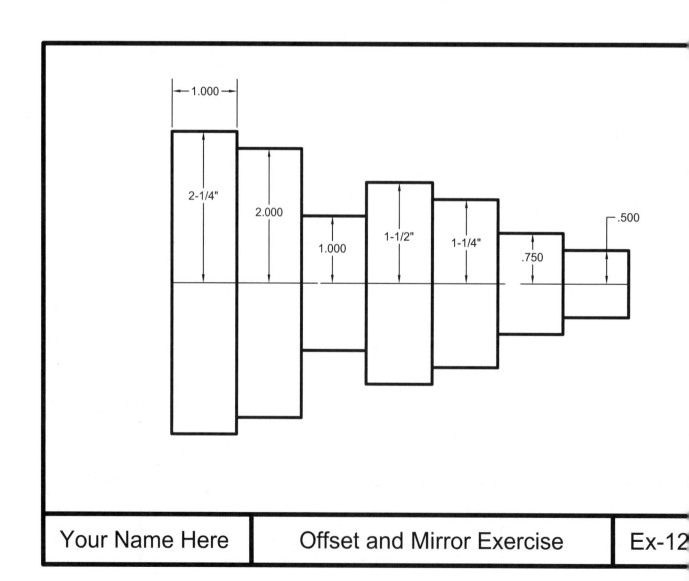

| Your Name Here | Offset and Mirror Exercise | Ex-12 |

EXERCISE 12D

INSTRUCTIONS:

1. Open **Border A**
2. Draw the lamp base below.
3. Practice using the "Layer method" when offsetting the hidden lines. Or use the Properties Palette or Quick Properties.
4. Notice that the part is symmetrical. Consider the Mirror command.
5. Use Layers, Object line, Hidden line and Centerline.
6. Edit the Title and Ex-XX by double clicking on the text. Do not erase and replace.
7. Do not dimension
8. Save as **EX12D**
9. Plot using Page Setup **Class Model A**

ISOMETRIC VIEW to help you visualize the object. DO NOT DRAW

| Your Name Here | Lamp Base | Ex-12D |

Notes:

LEARNING OBJECTIVES

After completing this lesson, you will be able to:

1. Create multiple copies in a rectangular or circular pattern
2. Understand how to Array objects

LESSON 13

ARRAY

The ARRAY command allows you to make multiple copies in a **RECTANGULAR** or Circular **(POLAR)** pattern. The maximum limit of copies per array is 100,000. This limit can be changed but should accommodate most users. (Refer to Help menu)

RECTANGULAR ARRAY

This method allows you to make multiple copies of object(s) in a rectangular pattern. You specify the number of rows (horizontal), columns (vertical) and the offset distance between the rows and columns. The offset distances will be equally spaced.

Offset Distance is sometimes tricky to understand. *Read this carefully*. The offset distance is the <u>distance from a specific location on the original to that same location</u> on the future copy. It is <u>not just the space in between</u> the two. Refer to the example below.

To use the rectangular array command you will select the object(s), specify how many rows and columns desired and the offset distance for the rows and the columns. **Refer to step by step instructions on page 13-3.**

Example of Rectangular Array:

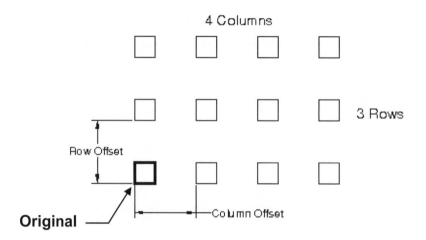

Example of Rectangular Array on an angle:
Notice the copies do not rotate. The Angle is only used to establish the placement.

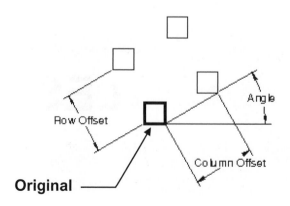

ARRAY....continued

RECTANGULAR ARRAY (Step by step)

1. Select the **ARRAY** command using one of the following:

    Ribbon = Home tab / Modify panel / ⊞

    Keyboard = Array <enter>

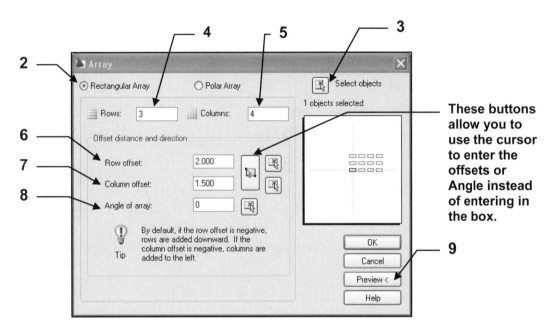

2. Select "**Rectangular Array**".

3. Select the **"Select Objects"** button.
    This will take you back to your drawing. Select the objects to Array then <enter>.

4. Enter the number of rows. (Horizontal rows)

5. Enter the number of columns. (Vertical columns)

6. Enter the row offset. (The distance from a specific location on the original to that same specific location on the future copy.) See example on pg. 13-2.

7. Enter the column offset. (The distance from a specific location on the original to that same specific location on the future copy.) See example on pg. 13-2.

8. Enter an angle if you would like the array to be on an angle.

9. Select the **Preview** button.
    If it looks correct, right click. If it is **not** correct, press the **ESC** key, make the necessary corrections and preview again.
    (Note: If the Preview button is gray, you have forgotten to "Select objects"- #3)

ARRAY....continued

POLAR ARRAY

This method allows you to make multiple copies in a <u>circular pattern</u>. You specify the total number of copies to fill a specific Angle or specify the angle between each copy and angle to fill.

To use the polar array command you select the object(s) to copy, specify the center of the array, specify the number of copies or the angle between the copies, the angle to fill and if you would like the copies to rotate as they are copied.

Example of Polar Array

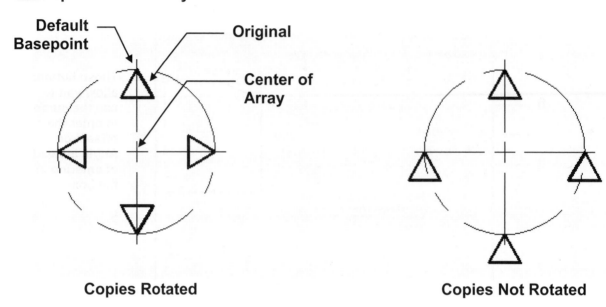

Default Basepoint   **Original**

Center of Array

Copies Rotated   **Copies Not Rotated**

Select "More" button to specify Objects Base Point (Refer to page 13-5)

Rotate Option

ARRAY....continued

Note: the two examples shown on the previous page use the **objects default base point**. The examples below displays what happens if you specify a basepoint.

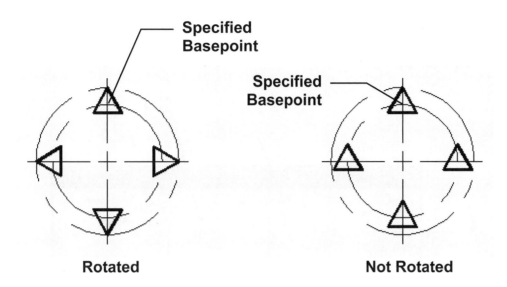

Rotated   **Not Rotated**

List of Default Base Point locations
Arc, Circle and Ellipse = Center of Object
Polygon and Rectangle = First Corner
Line, Polyline and Donut = Starting Point
Text and Block = Insertion Point

Note: If you select multiple objects the base point of the last object selected is used to construct the Array.

The difference between "Center Point" and "Object Base Point" is sometimes confusing.

Center Point
When specifying the Center Point, try to visualize the copies already there. Now, in your mind, try to visualize the center of that array. In other words, it is the Pivot Point from which the copies will be placed around.

Base Point
The Base Point is different. It is located on the original object. You must select the Base Point in addition to the Center Point. They are two different options.

ARRAY....continued

POLAR ARRAY (Step by step)

1. Select the ARRAY command using one of the following:

   Ribbon = Home tab / Modify panel / ⊞

   Keyboard = Array <enter>

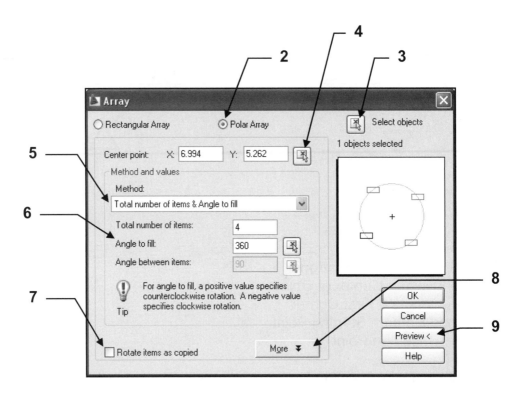

2. Select **" Polar Array".**

3. Select the **"Select Objects"** button.
   This will take you back to your drawing. Select the objects to Array then <enter>.

4. Select the "Center Point" button and select the center point with the cursor or enter the X and Y coordinates in the "Center Point" boxes.

5. Select the method. (Examples on the next page)

6. Enter the "Total number of items", "Angle to fill" or "Angle between items".

7. Select whether you want the items rotated as copied or not.

8. Accept the object's default base point or enter the X and Y coordinates.
   (Select the "More" button to show this area. Select the "Less" button to not show.)

9. Select the Preview button.
   <u>If it looks correct,</u> right click. <u>If it is **not** correct,</u> press the ESC key, make the necessary corrections and preview again.
   (Note: If the Preview button is gray, you have forgotten to "Select objects"- #3)

ARRAY....continued

Total number of items & Angle to fill

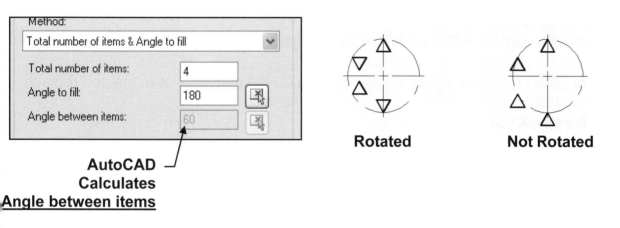

AutoCAD Calculates Angle between items

Rotated   Not Rotated

Total number of items & Angle between items

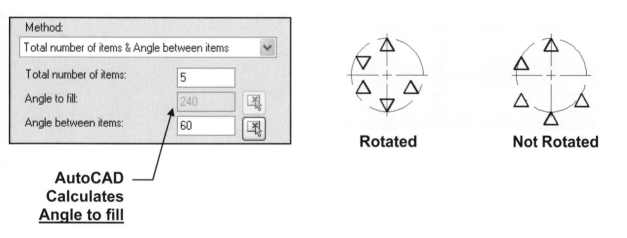

AutoCAD Calculates Angle to fill

Rotated   Not Rotated

Angle to fill & Angle between items

AutoCAD Calculates Number of items

Rotated   Not Rotated

EXERCISE 13A

INSTRUCTIONS:
1. Open **Border A**
2. Draw the original .500 square on Layer Object.
3. Array the original square as shown.
4. Edit the Title and Ex-XX by double clicking on the text. Do not erase and replace.
5. Do not dimension
6. Save as **EX13A**
7. Plot using Page Setup **Class Model A**

| Your Name Here | Rectangular Array | Ex-13 |

EXERCISE 13B

INSTRUCTIONS:
1. Open **Border A**
2. Draw the original .500 square on Layer Object.
3. Array the original square as shown. The Angle is 20
4. Edit the Title and Ex-XX by double clicking on the text. Do not erase and replace.
5. Do not dimension
6. Save as **EX13B**
7. Plot using Page Setup **Class Model A**

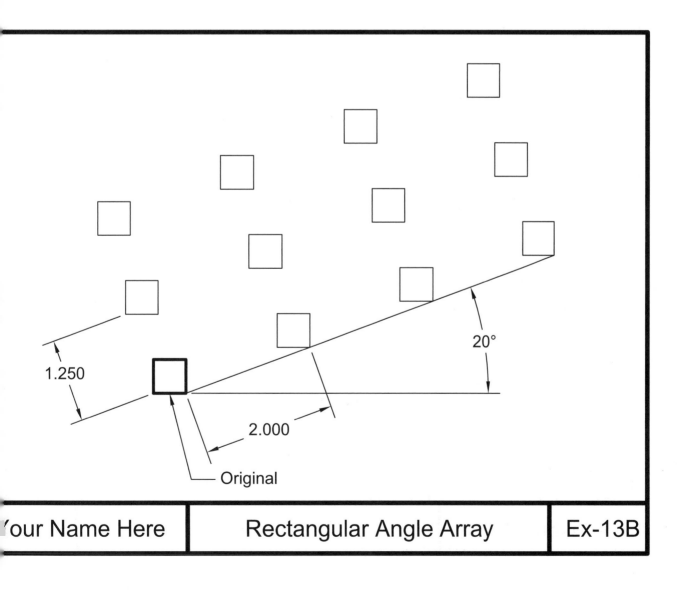

| Your Name Here | Rectangular Angle Array | Ex-13B |

EXERCISE 13C

INSTRUCTIONS:
1. Open **Border A**
2. Draw the objects shown below using the most efficient methods.
3. **Refer to the next page for more dimensions and helpful hints.**
4. Use Layers = Object line.
5. Edit the Title and Ex-XX by double clicking on the text. Do not erase and replace.
6. Do not dimension
7. Save as **EX13C**
8. Plot using Page Setup **Class Model A**

Refer to the next page for helpful hints and more dimensions.

| Your Name Here | Place mat | Ex-13 |

EXERCISE 13C
Helper

Notches

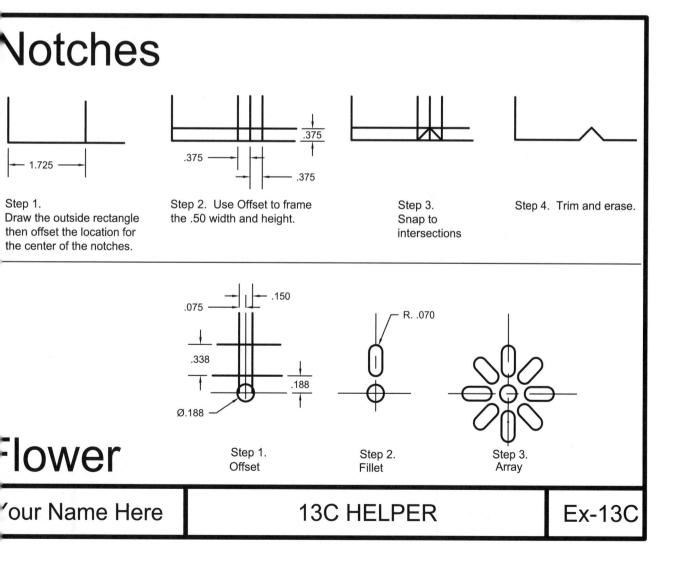

1.725

.375   .375   .375

Step 1.
Draw the outside rectangle
then offset the location for
the center of the notches.

Step 2. Use Offset to frame
the .50 width and height.

Step 3.
Snap to
intersections

Step 4. Trim and erase.

.075   .150

.338   .188

Ø.188

R. .070

Flower

Step 1.
Offset

Step 2.
Fillet

Step 3.
Array

Your Name Here	13C HELPER	Ex-13C

EXERCISE 13D

INSTRUCTIONS:

1. Open **Border A**
2. Draw the centerlines and circles first.
3. Draw one Polygon at the 12:00 location.
4. Array the original Polygon as shown.
5. Edit the Title and Ex-XX by double clicking on the text. Do not erase and replace.
6. Do not dimension
7. Save as **EX13D**
8. Plot using Page Setup **Class Model A**

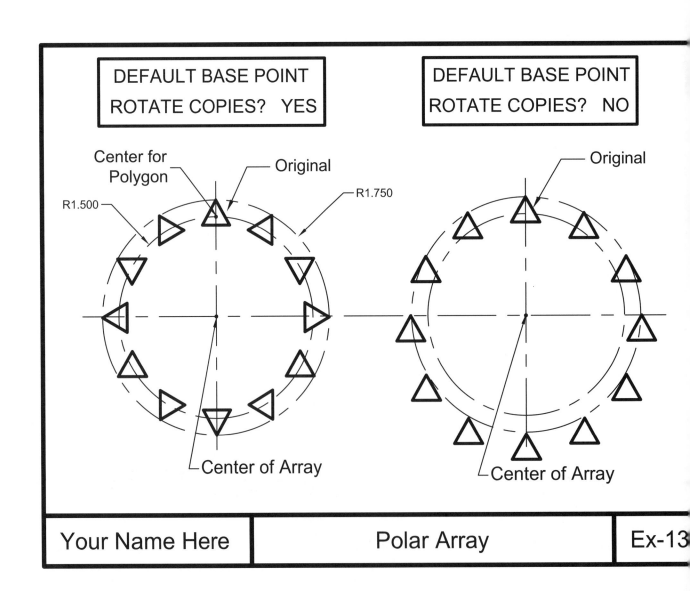

DEFAULT BASE POINT
ROTATE COPIES? YES

DEFAULT BASE POINT
ROTATE COPIES? NO

Center for Polygon

Original

R1.500

R1.750

Center of Array

Original

Center of Array

| Your Name Here | Polar Array | Ex-13 |

LEARNING OBJECTIVES

After completing this lesson, you will be able to:

1. Make an existing object larger or smaller proportionately
2. Stretch or compress an existing object.
3. Rotate an existing object to a specific angle.

LESSON 14

SCALE

The **SCALE** command is used to make objects larger or smaller <u>proportionately</u>. You may scale using a scale factor or a reference length. You must also specify a base point. The base point is a stationary point from which the objects scale.

1. Select the SCALE command using one of the following:

   Ribbon = Home tab / Modify panel / [icon]

   Keyboard = SCALE <enter>

SCALE FACTOR

    Command: _scale
2. Select objects: ***select the object(s) to be scaled***
3. Select objects: ***select more object(s) or <enter> to stop***
4. Specify base point: ***select the stationary point on the object***
5. Specify scale factor or [Copy/Reference]: ***type the <u>scale factor</u> <enter>***

If the scale factor is greater than 1, the objects will increase in size.
If the scale factor is less than 1, the objects will decrease in size.

REFERENCE option

Command: _scale
2. Select objects: ***select the object(s) to be scaled***
3. Select objects: ***select more object(s) or <enter> to stop***
4. Specify base point: ***select the stationary point on the object***
5. Specify scale factor or [Copy/Reference]: ***select Reference***
6. Specify reference length <1>: ***specify a <u>reference</u> length***
7. Specify new length: ***specify the new length***

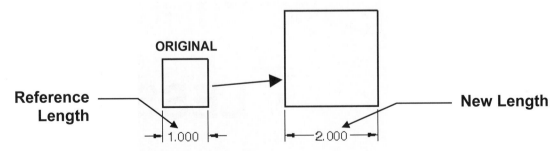

<u>**COPY**</u> **option** - creates a duplicate of the selected object. The duplicate is directly on top of the original. The duplicate will be scaled. The Original remains the same.

STRETCH

The **STRETCH** command allows you to stretch or compress object(s). Unlike the Scale command, you can alter an objects proportion with the Stretch command. In other words, you may increase the length without changing the width and vice versa.

Stretch is a very valuable tool. Take some time to really understand this command. It will save you hours when making corrections to drawings.

When selecting the object(s) you must use a **CROSSING** window.
Objects that are crossed, will ***stretch.***
Objects that are totally enclosed, will ***move***.

1. Select the STRETCH command using one of the following:

    Ribbon = Home tab / Modify panel /

    Keyboard = S <enter>

    Command: _stretch
2. Select objects to stretch by crossing-window or crossing-polygon...
    Select objects: ***select the first corner of the crossing window***
3. Specify opposite corner: ***specify the opposite corner of the crossing window***
4. Select objects: ***<enter>***
5. Specify base point or [Displacement] <Displacement>:
    select a base point (where it stretches from)
6. Specify second point or <use first point as displacement>:
    type coordinates or place location with cursor

Crossing Window (Placement of the crossing window is very important)

BEFORE

Base Point

AFTER (Notice which objects stretched and which objects moved.)

Second Point

ROTATE

The **ROTATE** command is used to rotate objects around a Base Point. (pivot point)

After selecting the objects and the base point, you will enter the rotation angle from its <u>current</u> rotation angle or select a reference angle followed by the new angle.

A **Positive** rotation angle revolves the objects **Counter- Clockwise**.
A **Negative** rotation angle revolves the objects **Clockwise**.

Select the ROTATE command using one of the following:

Ribbon = Home tab / Modify panel /

Keyboard = RO <enter>

ROTATION ANGLE OPTION
Command: _rotate
1. Current positive angle in UCS: ANGDIR=counterclockwise ANGBASE=0
   Select objects: *select the object to rotate.*
2. Select objects: *select more object(s) or <enter> to stop.*
3. Specify base point: *select the base point (pivot point).*
4. Specify rotation angle or [Copy/Reference]<0>: *type the angle of rotation.*

3. **Base point** — BEFORE — AFTER — 4. **Angle of Rotation** — 45°

REFERENCE OPTION
Command: _rotate
1. Current positive angle in UCS: ANGDIR=counterclockwise ANGBASE=0
   Select objects: *select the object to rotate.*
2. Select objects: *select more object(s) or <enter> to stop.*
3. Specify base point: *select the base point (pivot point).*
4. Specify rotation angle or [Reference]: *select Reference.*
5. Specify the reference angle <0>: *Snap to the reference object (1) and (2).*
6. Specify the new angle or [Points]: *P <enter>.*
7. Specify first point: *select 1st endpoint of new angle*
8. Specify second point: *select 2nd endpoint of new angle*

7. 1st endpt of new angle
8. 2nd endpt of new angle
Base Point
5. 1st Ref point — BEFORE — 2nd Refpoint
Rotation angles match
AFTER

EXERCISE 14A

INSTRUCTIONS:

1. Open **Border A**
2. Draw both 2" Squares shown below labeled "Original". Use layer Object Line.
3. Scale the original on the left using "scale factor" method.
4. Scale the original on the right using "scale Reference" method.
5. Edit the Title and Ex-XX by double clicking on the text. Do not erase and replace.
6. Do not dimension
7. Save as **EX14A**
8. Plot using Page Setup **Class Model A**

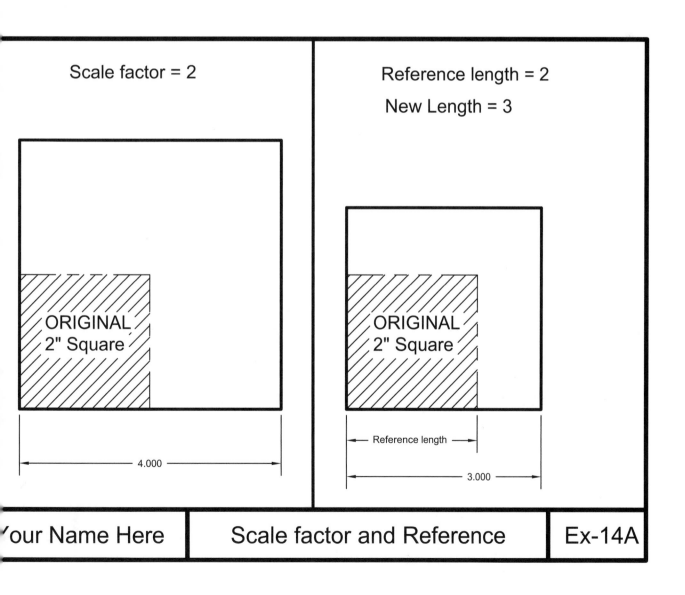

Scale factor = 2

Reference length = 2

New Length = 3

ORIGINAL
2" Square

ORIGINAL
2" Square

Reference length

4.000

3.000

Your Name Here | Scale factor and Reference | Ex-14A

EXERCISE 14B

INSTRUCTIONS:
1. Open **Border A**
2. Draw both 2" squares as shown below labeled as "Original". Use Layer Object Line
3. Stretch each as shown.
4. Edit the Title and Ex-XX by double clicking on the text. Do not erase and replace.
5. Do not dimension
6. Save as **EX14B**
7. Plot using Page Setup **Class Model A**

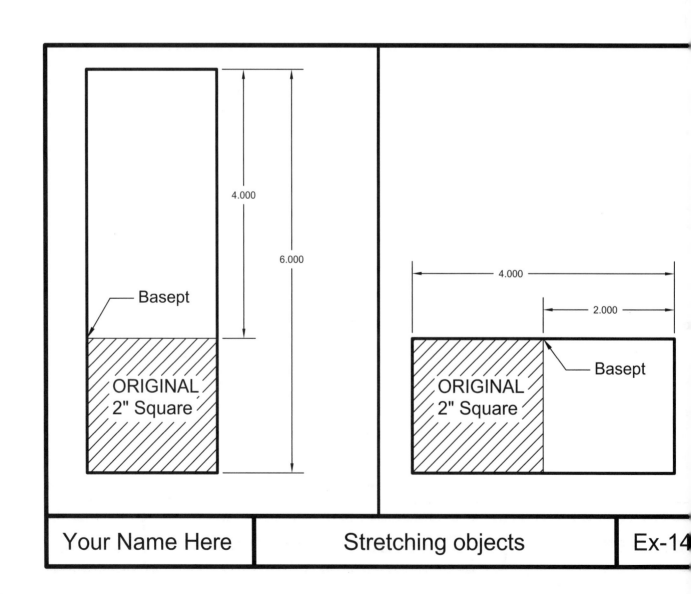

| Your Name Here | Stretching objects | Ex-14 |

EXERCISE 14C

INSTRUCTIONS:
1. Open **Border A**
2. Draw the Original shape on the left side of the border.
   Use Layer Object Line.

3. Stretch the overall length to 8.50 as shown.
   Notice the placement of the crossing window.

Crossing
Window

Stretch

4. Edit the Title and Ex-XX by double clicking on the text. Do not erase and replace.
5. Do not dimension
6. Save as **EX14C**
7. Plot using Page Setup **Class Model A**

Note:
If this was a floorplan consider how the stretch command could be very helpful.

EXERCISE 14D

INSTRUCTIONS:

1. Open **EX14C** — Crossing Window

2. Move the opening to the new location using the Stretch command.
   Notice the placement of the crossing window.

3. Edit the Title and Ex-XX by double clicking on the text. Do not erase and replace.
4. Do not dimension
5. Save as **EX14D**
6. Plot using Page Setup **Class Model A**

Note:
If this was a floorplan consider how the stretch command could be very helpful.

| Your Name Here | Moving a opening using Stretch | Ex-14 |

EXERCISE 14E

INSTRUCTIONS:
1. Open **Border A**
2. Draw the objects at position A. Use Layer Object Line.
3. Copy the objects to position B and C. (Shown as dashed lines)
4. Rotate the copies as shown. **Note**: Base point is important
5. Edit the Title and Ex-XX by double clicking on the text. Do not erase and replace.
6. Do not dimension
7. Save as **EX14E**
8. Plot using Page Setup **Class Model A**

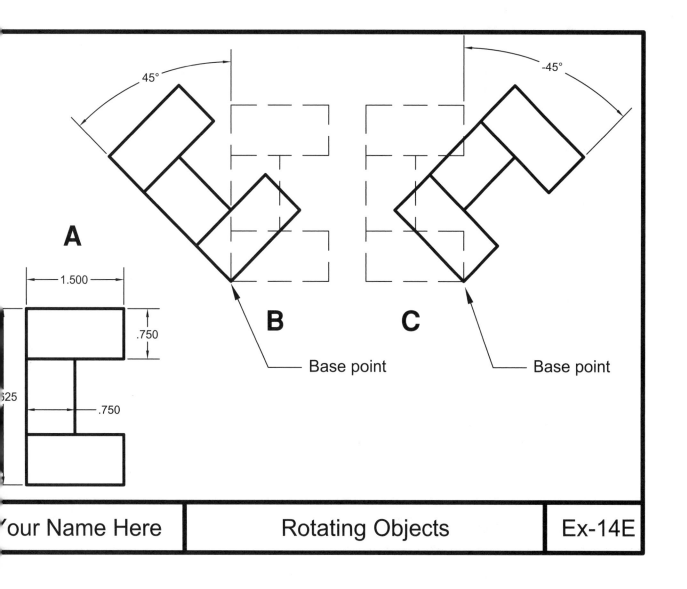

45°   -45°

A

1.500

.750

.750

625

B   C

Base point   Base point

| Your Name Here | Rotating Objects | Ex-14E |

NOTES:

After completing this lesson, you will be able to:

1. Place Cross hatch pattern in a boundary
2. Fill an area with solid fill
3. Add Gradient fills
4. Make changes to existing hatch sets.

LESSON 15

HATCH

The **HATCH** command is used to create hatch lines for section views or filling areas with specific patterns.

To draw **hatch** you must start with a closed boundary. A closed boundary is an area completely enclosed by objects. A rectangle would be a closed boundary. You simply pick inside the closed boundary. HATCH locates the area and automatically creates a temporary polyline around the outline of the hatch area. After the hatch lines are drawn in the area, the temporary polyline is automatically deleted. (Polylines are discussed in a future lesson)

Note: A Hatch set is one object. If you explode it, it will return to multiple objects such as lines.

It is good drawing management to always place gradient fill on it's own layer.
Use Layer Hatch

You may also make gradient fill appear or disappear with the **FILL** command

1. Select the Hatch command using one of the following:

   Ribbon = Home tab / Draw panel / ⊞

   Keyboard = BH or hatch <enter>

2. Select the hatch **Type**

   (**Predefined** and **User defined** described on pages 15-3 and 15-4)

2. **Select the Hatch type**

HATCH....continued

PREDEFINED

AutoCAD has many predefined hatch patterns.
These patterns are stored in the acad.pat and acadiso.pat files.
(You may also purchase patterns from other software companies.)

> **Note:** Using Hatch patterns will greatly increase the size of the drawing file. So use them conservatively.

To select a pattern by name, click on the "Pattern" down arrow. A drop-down list of available patterns will appear.

To select a pattern by appearance, click on the (...) button. This will display the Hatch Pattern Palette" dialog box.

This box displays the name of the Pattern you selected.

This determines the rotation angle of the pattern. A pre-designed pattern has a default angle of 0. If you change this Angle it will rotate the pattern relative to its original design.

The selected pattern is displayed here.

The value in this box is the scale factor. This will be discussed more later.

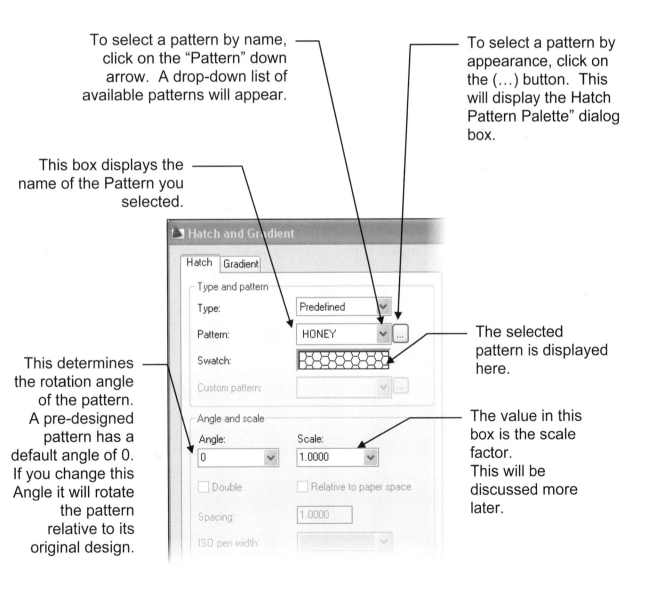

HATCH....continued

USER DEFINED

This selection allows you to simply draw continuous lines. (No special pattern)
You specify the **Angle** and the **Spacing** between the lines.

(This selection does not increase the size of the drawing file like Predefined)

Specify the
actual angle of
the hatch lines.
(0 to 180)

A sample of the
angle and spacing
settings is displayed
here.

Specify the actual
distance between
each hatch line.

*This option is only available when
using "User Defined")*
If this box is checked, the hatch set
will be drawn first at the angle
specified, then a second hatch set
will be drawn rotated 90 degrees
relative to the first hatch set, creating
a criss-cross affect.

HATCH....continued

3. Select the **Options**

Annotative: Automatically adjusts the scale to match the current Annotative scale.
This option will be discussed in Lesson 27.

Associative: The hatch set is one entity and if the boundary size is changed the hatch will automatically change to the new boundary shape.
(Refer to page 15-11 for an example)

Create separate hatches: Controls whether HATCH creates a single hatch object or separate hatch objects when selecting several closed boundaries.

4. Select the Area that you would like to Hatch.
   This operation is called "Selecting the boundary".

Add: Pick Points:
Select the **Pick Points** box then select a point inside the area you want to hatch. A boundary will automatically be determined.

Add: Select Objects:
Select the boundary by selecting the object(s).

Example of Pick Points:

Click inside boundary area.
The hatch will be placed inside the closed area.

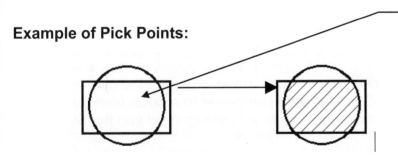

Example of Select Objects:

Select the Text and Rectangle
Hatch will be placed inside the Rectangle but outside the text.

HATCH....continued

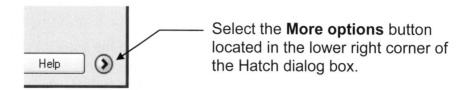

Select the **More options** button located in the lower right corner of the Hatch dialog box.

5. Boundary Retention
   It is important to select **Retain** or **Not retain** the boundary of the hatch.
   The hatch boundary can be retained, or disappear, if you erase the hatch lines.

6. If the area you selected to hatch is not completely closed (gaps) AutoCAD attempts to show you where the problem may have occurred. **Red circles** appear around endpoints near the gap and a warning appears.

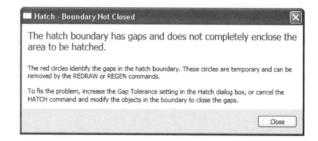

If a warning appears you may cancel the hatch command and close the gap or Increase the Gap Tolerance. The Gap tolerance can be set to a value from 0 to 5000. Any gaps equal to or smaller than the value you specify are ignored and the boundary is treated as closed.

HATCH....continued

7. Preview the Hatch

    a. After you have selected the boundary, press the right mouse button and select "**Preview**" from the short cut menu. This option allows you to preview the hatch set before it is actually applied to the drawing.

    b. If the preview is not what you expected, press the **ESC** key, make the changes and <u>preview again</u>.

    c. When you are satisfied, **right click** or press **<enter>** <u>to accept</u>.

       <u>Note: It is always a good idea to take the extra time to preview the hatch.</u>
                <u>It will actually save you time in the long run.</u>

HATCH....continued

HATCH - SOLID

If you would like to fill an area with a solid fill you should use **Predefined - Solid**.

> Note: The **Solid** pattern is a little difficult to find because it is not in alphabetical order. You will find it at the top of the pattern list.

You apply **Solid** hatch just like you would any other hatch pattern.
Select the pattern, select the boundary, preview and accept.

Examples of Solid hatch pattern:

GRADIENT FILL

Note: Gradient is not available in the LT version.

Gradients are fills that gradually change from dark to light or from one color to another. Gradient fills can be used to enhance presentation drawings, giving the appearance of light reflecting on an object, or creating interesting backgrounds for illustrations.

Gradients are definitely fun to experiment with but you will have to practice to achieve complete control.

1. Select the HATCH command using one of the following:

    Ribbon = Home tab / Draw panel /

    Keyboard = BH <enter>

2. Select the Gradient tab

3. Select **Boundary** using either Add: Pick Points or Select Objects.

4. Choose the Gradient settings described on the next page.

5. Preview

6. Accept or make changes.

GRADIENT FILL....continued

Color
Described below

9 Gradient Styles
Select one of the 9 gradient styles.
Note: The selected style will have a
white box around it.

CENTER
Select "**Centered**", with a check mark
in the box, to create a symmetrical fill.
Remove the check mark to change
the orientation of the fill.

ANGLE
Specify an angle for the highlighted
area from the drop down list.
(The angle rotates counter clockwise,
0 is upper left corner.)

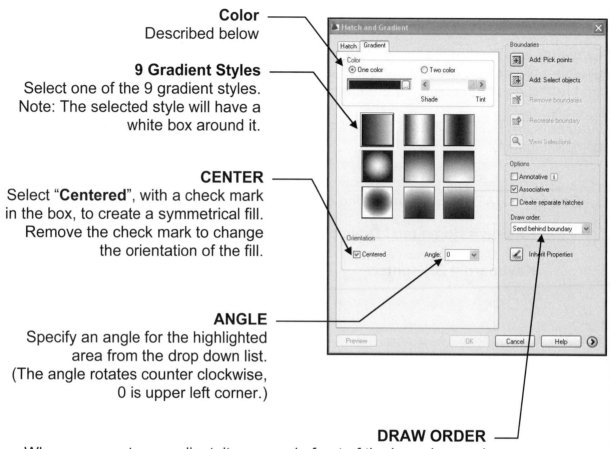

DRAW ORDER
When you create a gradient, it appears in front of the boundary and
sometimes obscures the object's outline. To send the gradient hatch
behind the boundary, specify the Draw Order.

ONE COLOR
Click the (...) button to the right of the color
swatch to open the Select Color dialog box.
Choose the color you want.

Use the "**Shade and Tint**" slider to choose
the gradient range from lighter to darker.

TWO COLOR
Click each (...) button to choose a color.
When you choose two colors the transition is
both from light to dark and from the first color
to the second.

EDITING HATCH

HATCHEDIT allows you to edit an existing hatch pattern in the drawing.
You simply double click on the hatch set that you want to change or type
he <enter> and select the hatch set you want to change and the hatch dialog box will
appear. Make the changes, preview and accept.

CHANGING THE BOUNDARY
If you change the shape of the boundary the hatch will conform to the new shape
if the hatch is ASSOCIATIVE. This means the hatch is associated to the object.
(Refer to 15-5)

Original Boundary   **New Boundary**   **New Boundary**
                          with Associative hatch   **with non-Associative hatch**

TRIMMING HATCH
You may trim a hatch set just like any other object.

EXERCISE 15A

INSTRUCTIONS:
1. Open **Border A**
2. Draw the Rectangle, Circle and Polygon as shown.
3. Use Layers = Object line and Hatch.
4. Hatch the boundaries shown using the Predefined Patterns shown.
5. Edit the Title and Ex-XX by double clicking on the text. Do not erase and replace.
6. Do not dimension
7. Save as **EX15A**
8. Plot using Page Setup **Class Model A**

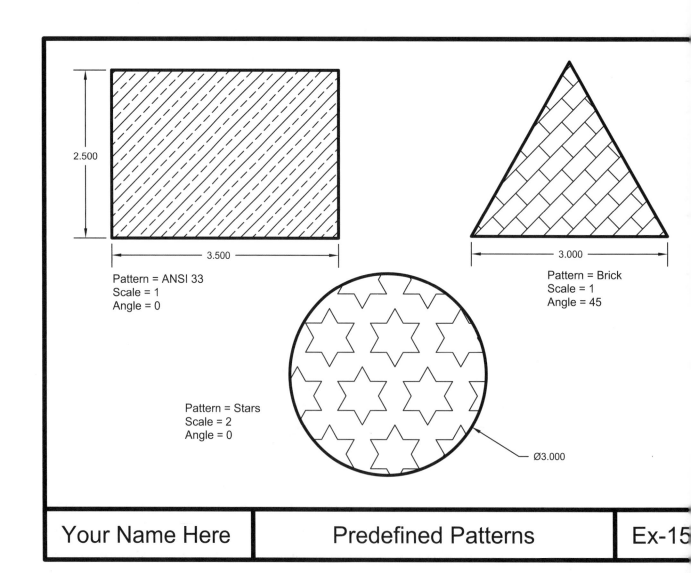

Pattern = ANSI 33
Scale = 1
Angle = 0

Pattern = Brick
Scale = 1
Angle = 45

Pattern = Stars
Scale = 2
Angle = 0

2.500

3.500

3.000

Ø3.000

Your Name Here	Predefined Patterns	Ex-15

EXERCISE 15B

INSTRUCTIONS:
1. Open **Border A**
2. Draw the 4 Rectangle as shown.
3. Use Layers = Object line and Hatch.
4. Hatch the boundaries shown using the Predefined Patterns shown.
   Notice the angles rotate the hatch pattern from it's original design.
   For example: Angle 45 appears to be rotation 90 because the original design
   was drawn at a 45 degree.....think about it.

5. Edit the Title and Ex-XX by double clicking on the text. Do not erase and replace.
6. Do not dimension
7. Save as **EX15B**
8. Plot using Page Setup **Class Model A**

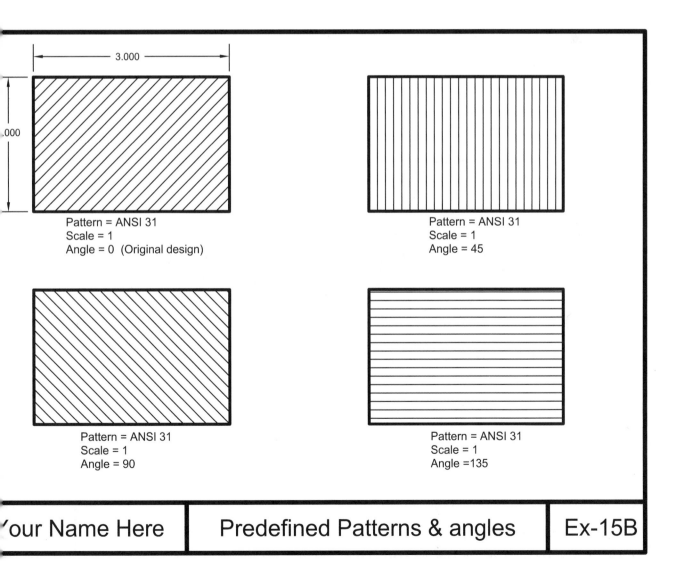

Pattern = ANSI 31
Scale = 1
Angle = 0 (Original design)

Pattern = ANSI 31
Scale = 1
Angle = 45

Pattern = ANSI 31
Scale = 1
Angle = 90

Pattern = ANSI 31
Scale = 1
Angle =135

| Your Name Here | Predefined Patterns & angles | Ex-15B |

EXERCISE 15C

INSTRUCTIONS:

1. Open **Border A**
2. Draw the Rectangle as shown using layer Object line.
3. Place the 1" high text in the Middle of the rectangle. Use layer Text.
4. Hatch using "Select Objects" this time. Use layer Hatch.
5. Edit the Title and Ex-XX by double clicking on the text. Do not erase and replace.
6. Do not dimension
7. Save as **EX15C**
8. Plot using Page Setup **Class Model A**

EXERCISE 15D

INSTRUCTIONS:

1. Open **Border A**
2. Draw the Circle as shown using layer Object line.
3. Using "User Defined" this time, hatch the circle with Double hatch. (See 15-4)
4. Hatch = Spacing = .50   Angle = 45   Use layer Hatch.
5. Edit the Title and Ex-XX by double clicking on the text. Do not erase and replace.
6. Do not dimension
7. Save as **EX15D**
8. Plot using Page Setup **Class Model A**

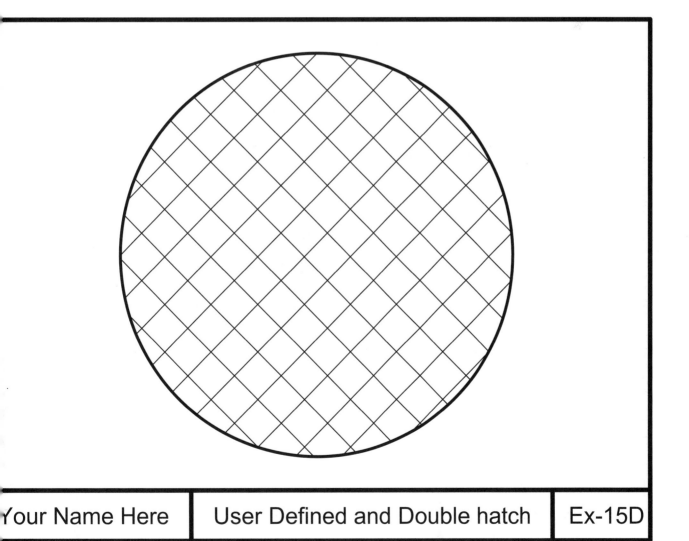

| Your Name Here | User Defined and Double hatch | Ex-15D |

EXERCISE 15E

INSTRUCTIONS:
1. Open **Border A**
2. Draw the Hub as shown using layer Object line.
3. Use Hatch pattern ANSI 31 Scale = 1 and Angle = 0.   Use Layer Hatch
4. Draw the section arrowhead, approximately as shown. Fill with "Solid" hatch.
5. Edit the Title and Ex-XX by double clicking on the text. Do not erase and replace.
6. Do not dimension
7. Save as **EX15E**
8. Plot using Page Setup **Class Model A**

| Your Name Here | HUB | Ex-15 |

EXERCISE 15F

INSTRUCTIONS:

1. Open **Border A**
2. Draw the objects shown as Before, using layer Object line.
3. Add Solid Hatch.   Use Layer Hatch
4. Edit the Title and Ex-XX by double clicking on the text. Do not erase and replace.
5. Do not dimension
6. Save as **EX15F**
7. Plot using Page Setup **Class Model A**

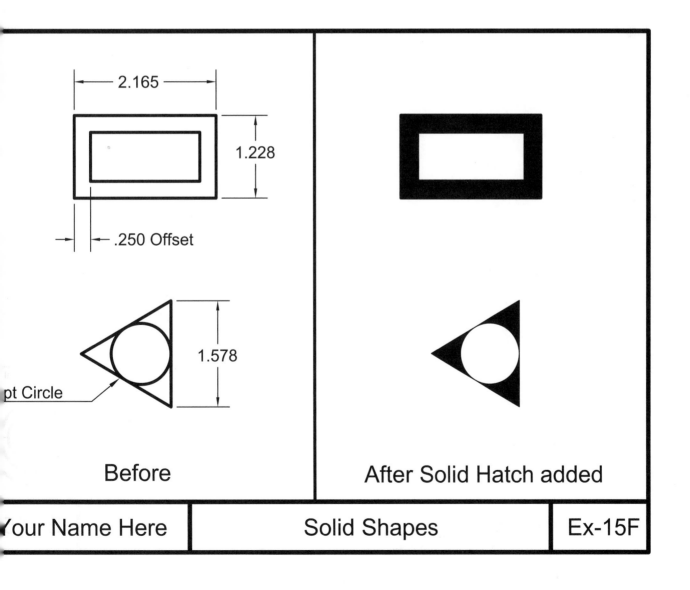

Before

After Solid Hatch added

| Your Name Here | Solid Shapes | Ex-15F |

Notes:

LEARNING OBJECTIVES

After completing this lesson, you will be able to:

1. Understand the importance of True Associative dimensioning
2. Create Linear, Continue and Baseline dimensions
3. Draw, Select and Erase objects
4. Use Grips to manipulate dimensions
5. Create a New dimension Style
6. Ignore Hatch Lines when dimensioning

LESSON 16

DIMENSIONING

*Dimensioning is basically easy, but as always, there are many options to learn.
As a result, I have divided the dimensioning process into multiple lessons.*

*In addition, Annotative dimensioning will be discussed in Lesson 27.
So relax and just take it one lesson at a time.*

In this Lesson you will learn how to create a **dimension style** and how to create Linear (horizontal and vertical) dimensions.

Dimensions can be Associative, Non-Associative or Exploded. You need to understand what these are so you may decide which setting you want to use. Most of the time you will use Associative but you may have reasons to also use Non-Associative and Exploded.

Associative
Associative Dimensioning means that the dimension is actually associated to the objects that they dimension. If you move the object, the dimension will move with it.
If you change the size of the object, the dimension text value will change also.
(Note: This is not parametric. In other words, you cannot change the dimension text value and expect the object to change. That would be parametric dimensioning)

Non-Associative
Non-Associative means the dimension is not associated to the objects and will not change if the size of the object changes.

Exploded
Exploded means the dimension will be exploded into lines, text and arrowsheads and non-associative.

How to set dimensioning to Associative, Non-Associative or Exploded.

1. On the command line type: ***dimassoc <enter>***

2. Enter the number ***2, 1 or 0 <enter>.***

   2 = Associative

   1 = Non-Associative

   0 = Exploded

DIMENSIONING....continued

How to Re-associate a dimension.

If a dimension is <u>Non-associative</u>, and you would like to make it <u>Associative</u>, you may use the **dimreassociate** command to change.

1. Select **Reassociate** using one of the following:

   Ribbon = Annotate tab / Dimension ▾ panel /

   Keyboard = Dimreassociate <enter>

2. Select objects: *select the dimension to be reassociated.*

3. Select objects: *select more dimensions or <enter> to stop.*

4. Specify first extension line origin or [Select object] <next>: (*an "X" will appear to identify which is the first extension); use object snap to select the exact location, on the object, for the extension line point.*

5. Specify second extension line origin <next>: (*the "X" will appear on the second extension) use object snap to select the exact location, on the object, for the extension line point.*

6. *Continue until all extension line points are selected.*

<u>Note: You must use object snap to specify the exact location for the extension lines.</u>

Regenerating Associative dimensions

Sometimes after panning and zooming, the associative dimensions seem to be floating or not following the object. The **DIMREGEN** command will move the associative dimensions back into their correct location.

Type: dimregen <enter>

LINEAR DIMENSIONING

Linear dimensioning allows you to create horizontal and vertical dimensions.

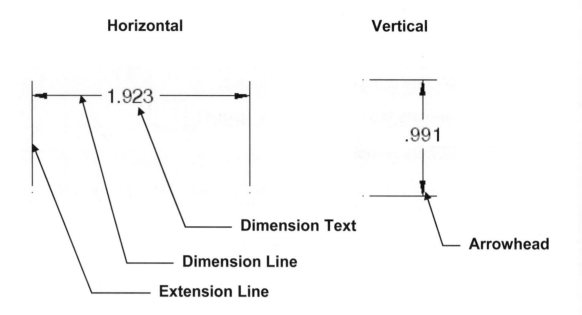

1. Select the **LINEAR** command using one of the following:

    Ribbon = Annotate tab / Dimension panel / [⊢]

    Keyboard = Dimlinear <enter>

    Command:__dimlinear
2. Specify first extension line origin or <select object>: ***snap to first extension line origin (P1).***
3. Specify second extension line origin: ***snap to second extension line origin (P2).***

4. Specify dimension line location or [Mtext/Text/Angle/Horizontal/Vertical/Rotated]: ***select where you want the dimension line placed (P3).***

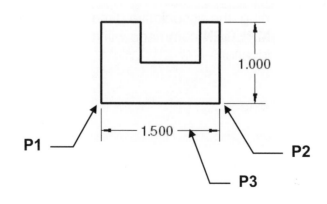

CONTINUE DIMENSIONING

Continue creates a series of dimensions in-line with an existing dimension. If you use the continue dimensioning immediately after a Linear dimension, you do not have to specify the continue extension origin.

1. Create a linear dimension first (**P1 and P2 shown below**)

2. Select the Continue command using one of the following:

   Ribbon = Annotate tab / Dimension panel / ⊞

   Keyboard = Dimcontinue <enter>

   Command: _dimcontinue
3. Specify a second extension line origin or [Undo/Select] <Select>: *snap to the second extension line origin (P3).*

4. Specify a second extension line origin or [Undo/Select] <Select>: *snap to the second extension line origin (P4).*

5. Specify a second extension line origin or [Undo/Select] <Select>: *press <enter> twice to stop.*

BASELINE DIMENSIONING

Baseline dimensioning allows you to establish a **baseline** for successive dimensions. The spacing between dimensions is automatic and should be set in the dimension style. (See 16-11, Baseline spacing setting)

A Baseline dimension must be used with an existing dimension. If you use Baseline dimensioning immediately after a Linear dimension, you do not have to specify the baseline origin.

1. Create a <u>linear</u> dimension first **(P1 and P2).**

2. Select the **BASELINE** command using one of the following:
   Command: _dimbaseline

   Baseline is in the drop-down menu with Continue.

   Ribbon = Annotate tab / Dimension panel / ⊟ ▼

   Keyboard = Dimbaseline <enter>

3. Specify a second extension line origin or [Undo/Select] <Select>: ***snap to the second extension line origin (P3).***

4. Specify a second extension line origin or [Undo/Select] <Select>: ***snap to P4.***

5. Specify a second extension line origin or [Undo/Select} <Select>: ***select <enter> twice to stop.***

Create this Linear dimension first

GRIPS

Grips are great tools for repositioning dimensions.
Grips are small, solid-filled squares that are displayed at strategic points on objects. You can drag these grips to stretch, move, rotate, scale, or mirror objects quickly.
Grips may be turned off by typing "grips" <enter> then 0 <enter> .

HOW TO USE GRIPS

1. Select the object (no command can be in use while using grips)

2. Select one of the **blue** grips. It will turn to "**red**". This indicates that grip is "**hot**".
   The "**Hot**" grip is the **basepoint**.

3. Move the hot grip to the new location.

4. <u>After editing you must press the **ESC** key to de-activate the grips.</u>

The following is an example of how to use grips to quickly reposition dimensions.

Grips may be used on all objects. If you like the example above, you should research grips further in the AutoCAD help menu.

DIMENSION STYLES

Using the "Dimension Style Manager" you can change the appearance of the dimension features, such as length of arrowheads, size of the dimension text, etc. There are over 70 different settings.
You can also Create New, Modify and Override Dimension Styles. All of these are simple, by using the Dimension Style Manager described below.

1. Select the "Dimension Style Manager" using one of the following:

Ribbon = Annotate tab / Dimension panel / ↘

Keyboard = Dimstyle <enter>

Displays which style is currently in use.

List all styles. **Annotative** will be discussed in Lesson 27.

Defined Below

Set Current Select a style from the list of styles and select the **set current** button.

New Select this button to create a new style. When you select this button, the **Create New Dimension Style** dialog box is displayed.

Modify Selecting this button opens the **Modify dimension Style** dialog box which allows you to make changes to the style selected from the "list of styles".

Override An override is a temporary change to the current style. Selecting this button opens the Override Current Style dialog box.

Compare Compares two styles.

CREATING A NEW DIMENSION STYLE

A dimension style is a group of settings that has been saved with a name you assign. When creating a new style you must start with an existing style, such as Standard. Next, assign it a new name, make the desired changes and when you select the OK button the new style will have been successfully created.

How to create a NEW dimension style.

1. Open **Border A**

2. Select the **Dimension Style Manager** command (Refer to previous page)

3. Select the **NEW** button.

4. Enter **CLASS STYLE** in the "**New Style Name**" box.

5. Select **STANDARD** in the "**Start With:**" box.

6. "**Use For:**" box will be discussed later. For now, leave it set to "**All dimensions**".

7. Select the **CONTINUE** button.

CREATING A NEW DIMENSION STYLE....continued

8. Select the **Primary Units** tab and change your setting to match the settings shown below.

8. **Primary Units**

Do not select the OK button yet.

CREATING A NEW DIMENSION STYLE....continued

9. Select the **Lines** tab and change your settings to match the settings shown below.

9. Lines

Do not select the OK button yet.

CREATING A NEW DIMENSION STYLE....continued

10. Select the **Symbols and Arrows** tab and change your setting to match the settings shown below.

10. Symbols and Arrows

Do not select the OK button yet.

CREATING A NEW DIMENSION STYLE....continued

11. Select the **Text** tab and change your settings to match the settings shown below.

11. Text

Do not select the OK button yet.

CREATING A NEW DIMENSION STYLE....continued

12. Select the **Fit** tab and change your setting to match the settings shown below.

13. Now select the OK button.

12. Fit

13. NOW select the OK button.

CREATING A NEW DIMENSION STYLE....continued

Your new style **"Class Style"** should be listed.

14. Select the **"Set Current"** button to make your new style "Class Style" the style that will be used.

14 ──────────────

New Style listed here

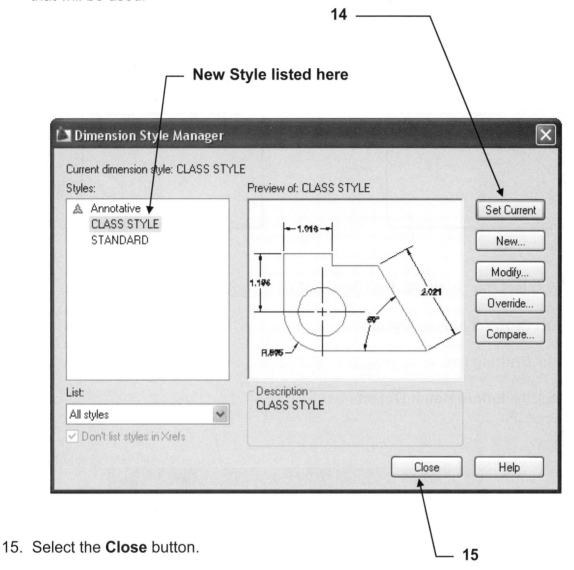

15. Select the **Close** button.

15

16. **Important:** Save your drawing as **Border A** again.

Note: You have successfully created a new "<u>Dimension</u> Style" called "**Class Style**". *This style will be saved <u>in your</u> **Border A** <u>drawing</u> after you save the drawing. It is important that you understand that this dimension style resides <u>only</u> in the **Border A** drawing. If you open another drawing, this dimension style will not be there.*

IGNORING HATCH OBJECTS

Occasionally, when you are dimensioning an object that has "Hatch Lines", your cursor will snap to the Hatch Line instead of the object that you want to dimension.
To prevent this from occurring, select the option **"IGNORE HATCH OBJECTS"**.

EXAMPLE:

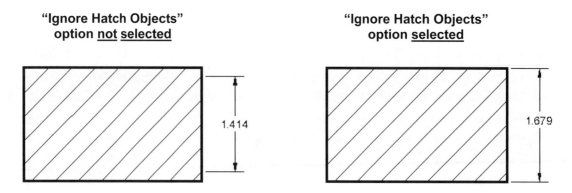

"Ignore Hatch Objects"
option <u>not</u> <u>selected</u>

1.414

"Ignore Hatch Objects"
option <u>selected</u>

1.679

<u>How to select the "IGNORE HATCH OBJECTS" option</u>

1. Select **Browser Menu / Options**

2. Select **Drafting** tab.

3. Check the **Ignore Hatch Objects** box.

EXERCISE 16A

INSTRUCTIONS:

1. Open **Border A**
2. Create a Dimension Style in your Border A by following the instructions on pages 16-9 through 16-15. Re-Save as Border A before starting the drawing shown below.
3. Draw the Objects as shown using layer Object line.
4. Dimension as shown using Dimension Style "Class Style" on Layer Dimension.
5. Use Linear, Continue and Baseline.
6. Edit the Title and Ex-XX by double clicking on the text. Do not erase and replace.
7. Save as **EX16A**
8. Plot using Page Setup **Class Model A**

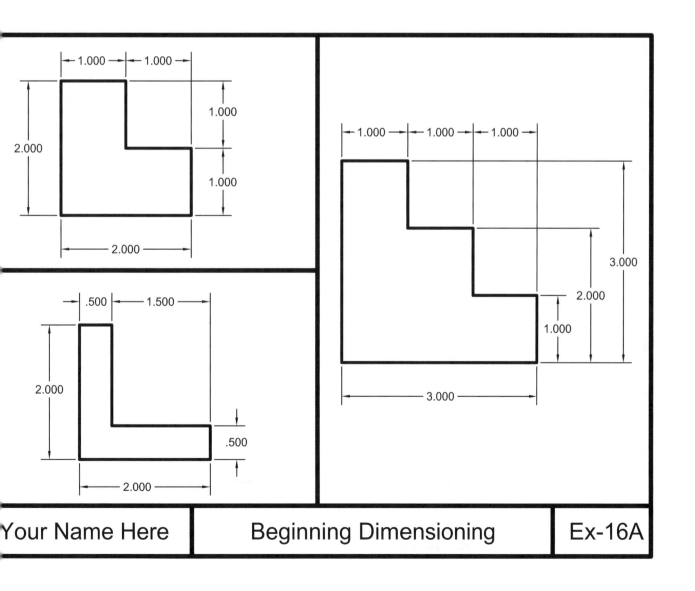

| Your Name Here | Beginning Dimensioning | Ex-16A |

EXERCISE 16B

INSTRUCTIONS:

1. Open **Border A**
2. Draw the Objects as shown using layer Object line.
3. Dimension as shown using Dimension Style "Class Style" on Layer Dimension.
4. Use Linear, Continue and Baseline.
5. Edit the Title and Ex-XX by double clicking on the text. Do not erase and replace.
6. Save as **EX16B**
7. Plot using Page Setup **Class Model A**

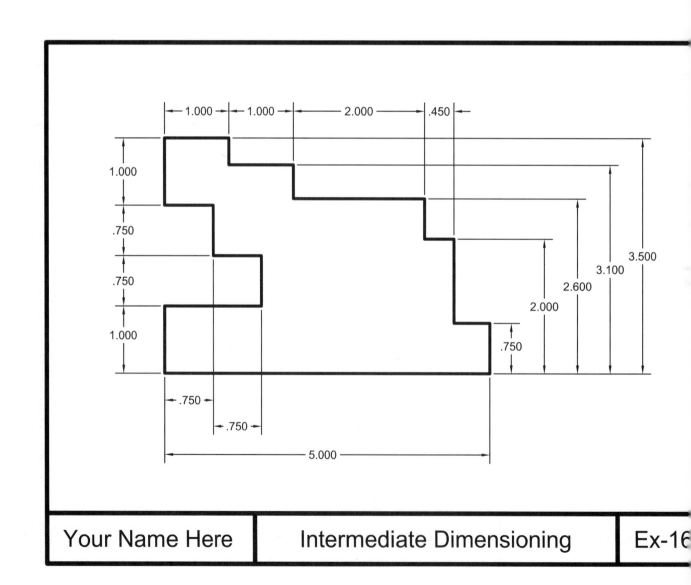

| Your Name Here | Intermediate Dimensioning | Ex-16 |

EXERCISE 16C

INSTRUCTIONS:

1. Open **Border A**
2. Follow the instructions below.
3. Do not divide your drawing into 4 sections. Just draw one rectangle in the middle of your drawing area and follow instructions.
4. Edit the Title and Ex-XX by double clicking on the text. Do not erase and replace.
5. Save as **EX16C**
6. Plot using Page Setup **Class Model A**

1. Draw a 2.250" long by 1.500" wide Rectangle.

2. Set "Dimassoc" setting to 2.
3. Dimension the Rectangle as shown here.

1.500

2.250

4. Stretch one corner of the rectangle using "Grips" approximately as shown (with Ortho and Osnap OFF)

1.500

2.250

Did the dimensions change?
Note: Your dimensions may not be exactly the same as shown. That's OK.

1.860

2.542

Your Name Here | Stretch and Associative Dims | Ex-16C

EXERCISE 16D

INSTRUCTIONS:
1. Open **Border A**
2. Follow the instructions below.
3. Do not divide your drawing into 4 sections. Just draw one rectangle in the middle of your drawing area and follow instructions.
4. Edit the Title and Ex-XX by double clicking on the text. Do not erase and replace.
5. Save as **EX16D**
6. Plot using Page Setup **Class Model A**)

1. Draw a 2.250" long by 1.500" wide Rectangle.

2. Set "Dimassoc" setting to 1. ◄— Note
3. Dimension the Rectangle as shown here.

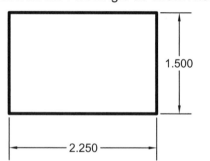

4. Stretch one corner of the rectangle using "Grips" approximately as shown (with Ortho and Osnap OFF)

Notice the difference?
The dimensions are not associated to the object, so they did not respond to the stretch.

| Your Name Here | Stretch and Non-Associative Dims | Ex-16 |

LEARNING OBJECTIVES

After completing this lesson, you will be able to:

1. Edit the dimension text without changing the value
2. Move the dimension text within the dimension lines
3. Modify an entire dimension style
4. Override a dimension style
5. Use the Properties Palette to change a dimension
6. Break intersecting extension and dimension lines
7. Add a Jog to a dimension line
8. Adjust the distance between dimensions

LESSON 17

EDITING DIMENSION TEXT VALUES

Sometimes you need to modify the dimension text value. You may add a symbol, a note or even change the text of an existing dimension.
The following describes 2 methods.

Example: Add the word "Max." to the existing dimension value text.

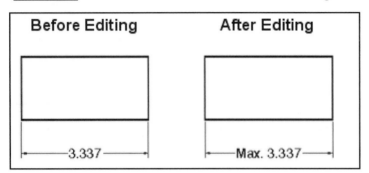

Method 1 (Properties Palette).
1. **Quick Properties** button must be **ON**.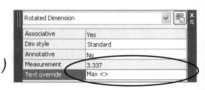
2. Select the dimension you want to override.
3. Scroll down to **Text override**
   (Notice that the actual measurement is directly above it.)
4. Type the new text (**Max**) and **< >** and press <enter>
   (**< >** represents the associative text value 3.337)

Important:
If you do not use < > the dimension will no longer be Associative.

Method 2 (Text Edit).
1. Type on the Command line: **ed** <enter>   (This is the **text edit** command)
2. Select the dimension you want to edit.

Associative Dimension
If the dimension is <u>Associative</u> the dimension text will appear highlighted.

 Before

You may add text in front or behind the dimension text and it will remain Associative. Be careful not to disturb the dimension value text.

— Max 9.815 — **After**

Non Associative or Exploded Dimension
If the dimension value has been changed or exploded it will appear with a gray background and is not Associative.

— 9.815 Max —

4. Make the change.
5. Select the **OK** button.

EDITING THE DIMENSION POSITION

Sometimes dimensions are too close and you would like to stagger the text or you need to move an entire dimension to a new location, such as the examples below.

Before Editing   Slide the text down   Move an entire dimension

Editing the position is easy.
1. Select the dimension that you want to change. (Grips will appear)
2. Select the middle grip. (It will become red and active, hot)
3. Move the cursor and the dimension will respond. (Osnap should be OFF)
4. Press the left mouse button to place the dimension in the new location.

(Refer to page 16-7 also for grips information)

ADDITIONAL EDITING OPTIONS USING THE SHORTCUT MENU

1. Select the dimension that you want to change.
2. Press the Right Mouse button.
3. Select "Dimension Text Position" from the shortcut menu shown to the right.
4. A sub-menu appears with many editing options.

MODIFY AN <u>ENTIRE</u> DIMENSION STYLE

After you have created a Dimension Style, you may find that you have changed your mind about some of the settings. You can easily change the entire Style by using the "Modify" button in the Dimension style Manager dialog box. This will not only change the Style for future use, but it will also <u>update dimensions already in the drawing</u>.

Note: if you do not want to update the dimensions already in the drawing, but want to make a change to a new dimension, refer to <u>Override</u>, page 17-5

1. Select the <u>Dimension Style Manager</u>. (Refer to page 16-8)

2. Select the Dimension Style that you wish to modify.

3. Select the **Modify button** from the Dimension Style Manager dialog box.

4. Make the desired changes to the settings.

5. Select the **OK** button.

6. Select the **Close** button.

Now look at your drawing. Have your dimensions updated?

Note:
The method above <u>will not</u> change dimensions that have been modified or exploded.

> **Note: If some of the dimensions have not changed:**
> 1. Type: -**Dimstyle <enter>**   (**notice the (-)dash infront of "dimstyle"**)
> 2. Type: **A <enter>**
> 3. Select dimensions to update and then <enter>
> (Sometimes you have to give them a little nudge.)

OVERRIDE A DIMENSION STYLE

A dimension Override is a **temporary** change to the dimension settings.
An override **will not affect existing dimensions**. It will **only** affect **new dimensions**.
Use this option when you want a **new** dimension just a little bit different but you don't
want to create a whole new dimension style and you don't want the existing
dimensions to change either.

For example, if you want the new dimension to have a text height of .500 but you want
the existing text to remain at .125 ht.

1. Select the **Dimension Style Manager.** (Refer to page 16-8)

2. Select the "**Style**" you want to override. (Such as: Class Style)

3. Select the **Override** button.

4. Make the desired changes to the settings. (Such as: Text ht = .500)

5. Select the **OK** button.

6. Confirm the Override
   Look at the List of styles.
   Under the Style name, a sub heading of **<style overrides>** should be displayed.
   The description box should display the style name and the override settings.

7. The description box should display the style name and the override settings.

8. Select the **Close** button.

When you want to return to style **Class Style**, select **Class Style** from the styles list
then select the **Set Current** button. Each time you select a different style, you must
select the **Set Current** button to activate it.

EDIT AN INDIVIDUAL EXISTING DIMENSION

Sometimes you would like to modify the settings of an **individual existing** dimension. This can be achieved using the **Properties palette**.

1. Double click on the dimension that you wish to change.

2. Select and change the desired settings.

    Example: Change the dimension text height to .500.

3. Press <enter> . *(The dimension should have changed)*

4. Close the Properties Palette.

5. Press the <esc> key to de-activate the grips.

Note: The dimension will remain Associative.

DIMENSION BREAKS

Occasionally extension lines overlap another extension line or even an object. If you do not like this you may use the **Dimbreak** command to break the intersecting lines. **Automatic** (described below) or **Manual** (described on page 17-8) method may be used. You may use the **Restore** (described on page 17-9) option to remove the break.

AUTOMATIC DIMENSION BREAKS

To create an automatically placed dimension break, you select a dimension and then use the Auto option of the DIMBREAK command.

Automatic dimension breaks are updated any time the dimension or intersecting objects are modified.

You control the size of automatically placed dimension breaks on the Symbols and Arrows tab of the Dimension Style dialog box.

The specified size is affected by the dimension break size, dimension scale, and current annotation scale for the current viewport. (Annotation discussed in Lesson 27)

1. Select the Dimbreak command using one of the following:

   Ribbon = Annotate tab / Dimension panel /

   Keyboard = Dimbreak <enter>

    Command: _Dimbreak
2. Select dimension to add / remove break or [Multiple]: **type M <enter>**

3. Select dimensions: **select dimension**

4. Select dimensions: **select another dimension**

5. Select dimensions: **select another dimension or <enter> to stop selecting**

6. Select object to break dimension or [Auto / Remove] <Auto>: **<enter>**

Before Dimbreak

After Dimbreak

DIMENSION BREAK....continued

MANUAL DIMENSION BREAK

You can place a dimension break by <u>picking two points</u> on the dimension, extension, or leader line to determine the size and placement of the break.

Dimension breaks that are added manually by picking two points are not automatically updated if the dimension or intersecting object is modified. So if a dimension with a manually added dimension break is moved or the intersecting object is modified, you might have to restore the dimension and then add the dimension break again.

The size of a dimension break that is created by picking two points is not affected by the current dimension scale or annotation scale value for the current viewport.
(You will learn Annotation in Lesson 27)

1. Select the Dimbreak command using one of the following:

   Ribbon = Annotate tab / Dimension panel /

   Keyboard = Dimbreak <enter>

Command: _Dimbreak
2. Select dimension to add / remove break or [Multiple]: *select the dimension*

3. Select object to break dimension or [Auto/Manual/Restore] <Auto>: *type M <enter>*

4. Specify first break point: *select the first break point location* (Osnap should be off)

5. Specify second break point: *select the second break point location*

Before Dimbreak   **After Dimbreak**

The following objects can be used as cutting edges when adding a dimension break:
Dimension, Leader, Line, Circle, Arc, Spline, Ellipse, Polyline, Text, and Multiline text.

DIMENSION BREAK....continued

REMOVE THE BREAK

Removing the break is easy using the <u>Restore</u> option.

1. Select the Dimbreak command using one of the following:

   Ribbon = Annotate tab / Dimension panel /

   Keyboard = Dimbreak <enter>

   Command: _DIMBREAK
2. Select dimension to add / remove break or [Multiple]: *type M <enter>*

3. Select dimensions: *select a dimension*

4. Select dimensions: *select a dimension*

5. Select dimensions: *select another dimension or <enter> to stop selecting*

6. Select object to break dimension or [Auto / Remove] <Auto>: *type R <enter>*

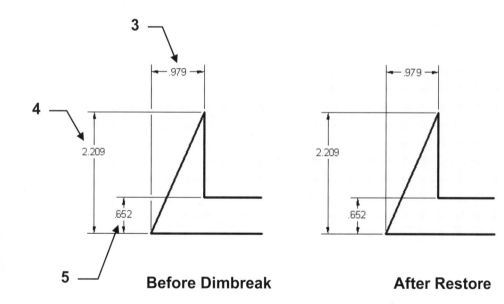

Before Dimbreak   **After Restore**

JOG A DIMENSION LINE

Jog lines can be added to linear dimensions. Jog lines are used to represent a dimension value that does not display the actual measurement.

Before you add a jog, the Jog angle and the height factor of the jog should be set in the Symbols & Arrows tab within the Dimension Style.

The height is calculated as a factor of the Text ht.

For example:
if the text height was .250 and the jog height factor was 1.500, the jog would be .375
Formula: (.250 ht. X 1.500 jog ht. factor= .375).

1. Select the Jogged Linear command

    Keyboard = Dimjogline <enter>

    Command: _DIMJOGLINE
2. Select dimension to add jog or [Remove]: *Select a dimension*

3. Specify jog location (or press ENTER): *Press <enter>*

4. <u>After you have added the jog you can position it by using grips and adjust the height of the jog symbol using the Properties palette.</u>

REMOVE A JOG
1. Select the Jogged Linear command
    Command: _DIMJOGLINE

2. Select dimension to add jog or [Remove]: *type R <enter>*

3. Select jog to remove: *select the dimension*

ADJUST DISTANCE BETWEEN DIMENSIONS

The Adjust Space command allows you to adjust the distance between existing parallel linear and angular dimensions, so they are equally spaced. You may also align the dimensions to create a string.

1. Select the Dimension Space command using one of the following:

   Ribbon = Annotate tab / Dimension panel /

   Keyboard = Dimspace <enter>

   Command: _DIMSPACE
2. Select base dimension: ***Select the dimension that you want to use as the base dimension when equally spacing dimensions. (P1)***
3. Select dimensions to space: ***select the next dimension to be spaced. (P2)***
4. Select dimensions to space: ***select the next dimension to be spaced. (P3)***
5. Select dimensions to space: ***continue selecting or press <enter> to stop***
6. Enter value or [Auto] <Auto>: ***enter a value or press <enter> for auto***

Auto creates a spacing value of twice the height of the dimension text.   For example: If the dimension text is 1/8", the spacing will be 1/4".

Aligning dimensions

Follow the steps shown above but when asked for the value enter **"0"** <enter>.

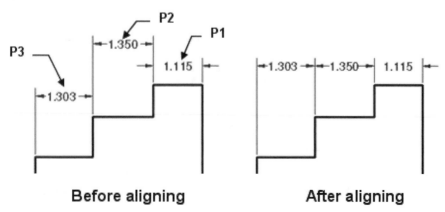

Before aligning   After aligning

EXERCISE 17A

INSTRUCTIONS:

1. Open **Border A**
2. Draw the 6" long by 4" wide Rectangle shown below. Use layer Object line.
3. **Upper Dimension:** Use dimension style Class Style . Use Layer Dimension
4. **Lower Dimension:** Use Override. Change the text height setting to .500.
   If the upper dimension changed also, you did not use override. Try again.
5. Edit the Title and Ex-XX by double clicking on the text. Do not erase and replace.
6. Save as **EX17A**
7. Plot using Page Setup **Class Model A**

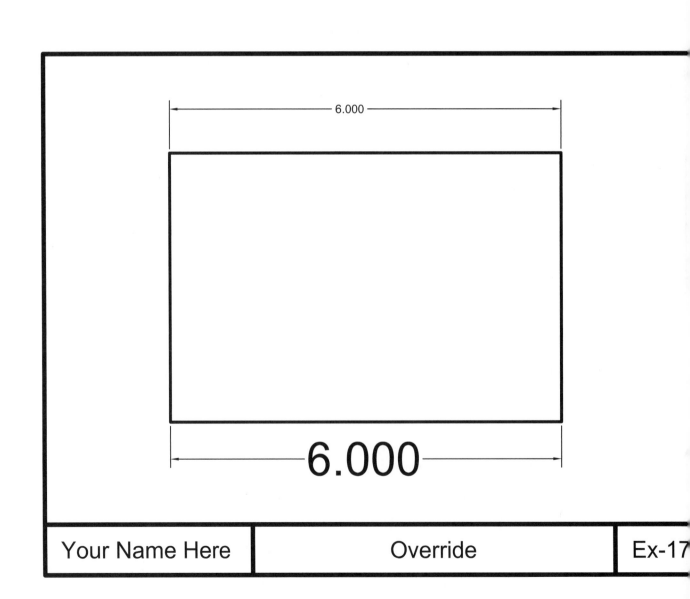

Your Name Here	Override	Ex-17

EXERCISE 17B

INSTRUCTIONS:

1. Open **EX16B**
2. Re-space the baseline dimensions on the right side. Use Auto for spacing.
3. Align the (2) .750 dimensions.
4. Edit the Title and Ex-XX by double clicking on the text. Do not erase and replace.
5. Save as **EX17B**
6. Plot using Page Setup **Class Model A**

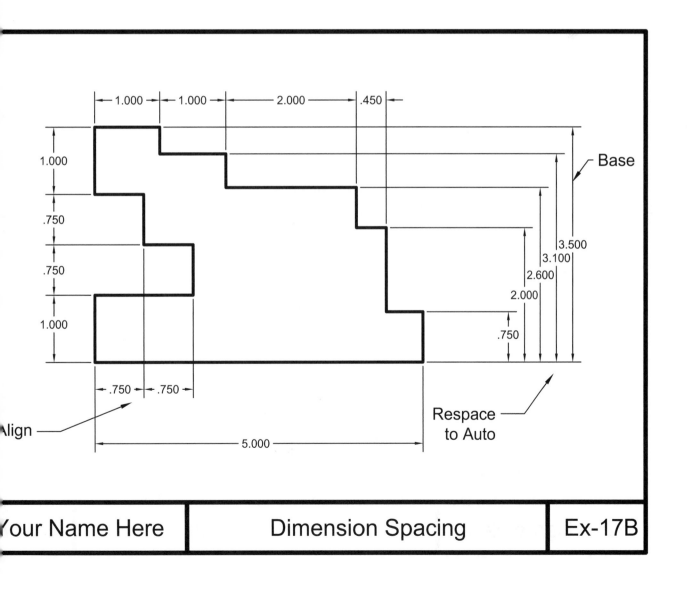

| Your Name Here | Dimension Spacing | Ex-17B |

EXERCISE 17C

INSTRUCTIONS:
1. Open **EX17B**
2. Make the changes shown below. Use Grips and the Properties Palette.
3. Notice: Not all dimensions change, so you can't make changes to the Dim. Style.
4. Edit the Title and Ex-XX by double clicking on the text. Do not erase and replace.
5. Save as **EX17C**
6. Plot using Page Setup **Class Model A**

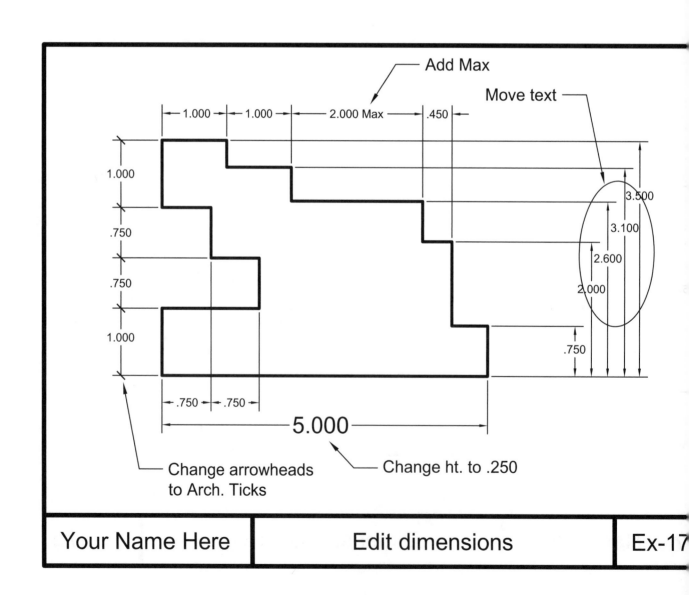

| Your Name Here | Edit dimensions | Ex-17 |

EXERCISE 17D

INSTRUCTIONS:

1. Open **Border A**
2. Draw the objects shown below.
3. Dimension as shown. Be very careful because there is just enough drawing area.
4. Use Dimension Break to break the extension lines as shown.
5. Use Dimension Jog to create a jog in the lower dimension.
6. Edit the Title and Ex-XX by double clicking on the text. Do not erase and replace.
7. Save as **EX17D**
8. Plot using Page Setup **Class Model A**

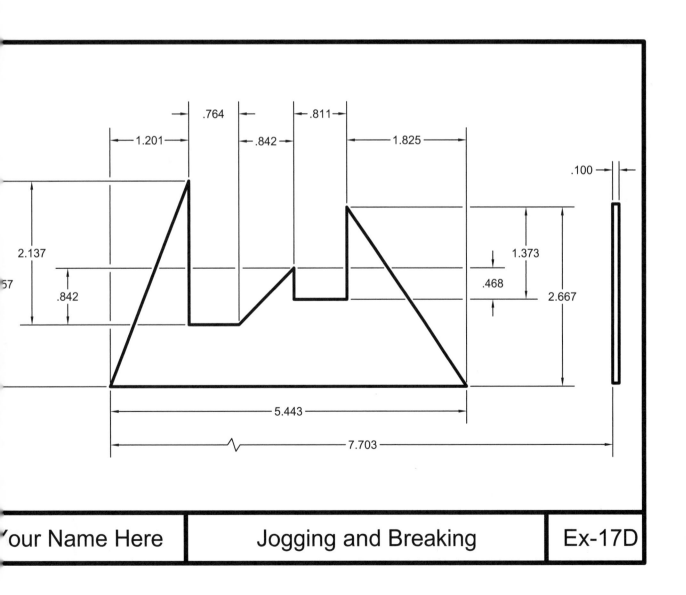

| Your Name Here | Jogging and Breaking | Ex-17D |

Notes:

LEARNING OBJECTIVES

After completing this lesson, you will be able to:

1. Add Diameter, Radius and Angular dimensions to a drawing
2. Draw and control Center Marks
3. Flip the direction of an arrowhead
4. Understand the need for Sub-styles
5. Create a sub-style

LESSON 18

DIMENSIONING DIAMETERS

The **DIAMETER** dimensioning command should be used when dimensioning circles and arcs of <u>more than 180 degrees</u>. AutoCAD measures the selected circle or arc and displa the dimension text with the diameter symbol (Ø) in front of it.

1. Select the Diameter command using one of the following:

    Ribbon = Annotate tab / Dimension panel /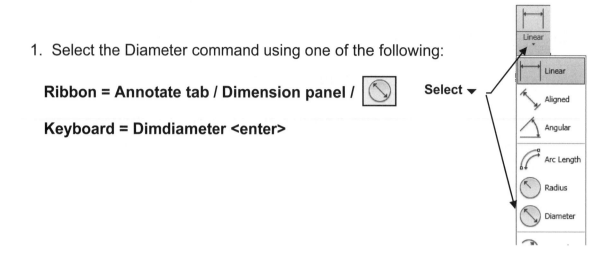

    Keyboard = Dimdiameter <enter>

2. Select arc or circle: *select the arc or circle (P1)* <u>*do not use object snap.*</u>
    Dimension text = *the diameter will be displayed here.*

3. Specify dimension line location or [Mtext/Text/Angle]: *place dimension text location (P*

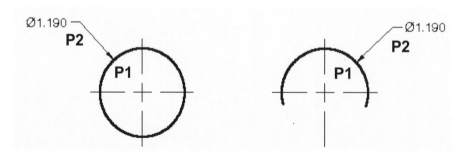

<u>Center marks</u> are automatically drawn as you use the diameter dimensioning command. If the circle already has a center mark or you do not want a center mark, set the center mark setting to **<u>NONE</u>** (Dimension Style / Symbols and Arrows tab) before using Diamet dimensioning. (Refer to 18-7)

DIMENSIONING DIAMETERS….continued

Controlling the diameter dimension appearance.

If you would like your Diameter dimensions to appear as shown in the two examples below, you must change some setting in the Dimension Style **Fit tab**.

DIMENSIONING RADII

The **Radius** dimensioning command should be used when dimensioning arcs of less tha
180 degrees. AutoCAD measures the selected arc and displays the dimension text with
the radius symbol (R) in front of it.

1. Select the Radius command using one of the following:

    Ribbon = Annotate tab / Dimension panel / **Select ▾**

    Keyboard = Dimradius <enter>

2. Select arc or circle: *select the arc (P1) do not use object snap.*
    Dimension text = *the radius will be displayed here.*

3. Specify dimension line location or [Mtext/Text/Angle]: *place dimension text location (P*

P2
R.500
P1

Center marks are automatically drawn as you use the radius dimensioning command.
If you do not want a center mark, set the center mark setting to **NONE**
(Dimension Style / Symbols and Arrows tab) before using Diameter dimensioning.
(Refer to 18-7)

DIMENSIONING RADII

Controlling the radius dimension appearance.

If you would like your Radius dimensions to appear as shown in the example below, you must change the "**Fit Options**" and "**Fine Tuning**" in the **Fit tab** in your Dimension Style.

ANGULAR DIMENSIONING

The **ANGULAR** dimension command is used to create an angular dimension between two lines that form an angle. AutoCAD determines the angle between the selected lines and displays the dimension text followed by a degree (°) symbol.

1. Select the **ANGULAR** command.

   Ribbon = Annotate tab / Dimension panel /   Select ▼

   Keyboard = Dimangular <enter>

2. Select arc, circle, line, or <specify vertex>: *select the first line that forms the angle (P1) location is not important, do not use object snap.*

3. Select second line: *select the second line that forms the angle (P2)*

4. Specify dimension arc line location or [Mtext/Text/Angle]: *place dimension text location (P3)*

   Dimension text = *angle will be displayed here*

   Any of the 4 angular dimensions shown below can be displayed by moving the cursor in the direction of the dimension after selecting the 2 lines (P1 and P2) that form the angle.

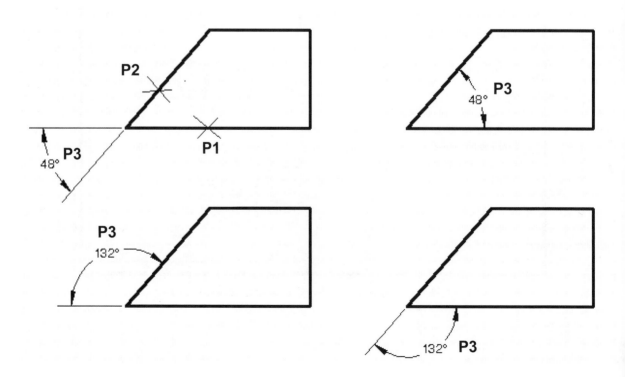

CENTERMARK

CENTERMARKS can ONLY be drawn with circular objects like Circles and Arcs.
You set the size and type.

The Center Mark has three <u>types</u>, **None, Mark** and **Line** as shown below.

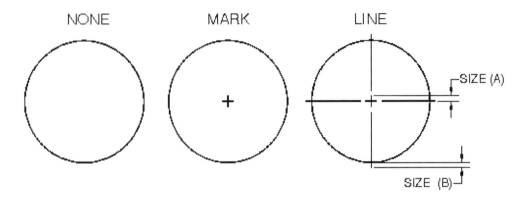

What does "SIZE" mean?

The **size** setting determines both, (A) the length of half of the intersection line and (B) the length extending beyond the circle. (See above right)

Where do you set the CENTERMARK "TYPE" and "SIZE"

1. Select the **Dimension Style** command.
2. Select: **Modify or Override.**
3. Select: **SYMBOLS and ARROWS** tab
4. Select the **Center mark type**
5. Set the **Size**.

To draw a CENTER MARK

1. Select the **Center mark** command using one of the following:

   Ribbon = Annotation tab / Dimension ▼ panel /

   Keyboard = DCE <enter>

2. Select arc or circle: ***select the arc or circle with the cursor.***

FLIP ARROW

You can easily **flip** the direction of the arrowhead using the **Flip Arrow** option.

How to <u>Flip an arrowhead</u>.

1. Select the arrowhead that you wish to Flip.

2. Press the right mouse button and select **Flip Arrow** from the menu.

3. The arrowhead Flips.

CREATING A DIMENSION SUB-STYLE

In Lesson 16 you learned how to create a Dimension Style named Class Style. All of the dimension created with that style appear identical because they have the same settings. Now you are going to learn how to create a "Sub-Style" of the Class Style.

For example:
If you wanted all of the **Diameter** dimensions to have a centerline automatically displayed but you did not want a centerline displayed when using the **Radius** dimension command. To achieve this, you must create a "sub-style" for all **Radius** dimensions.

Sub-styles have also been called "children" of the "Parent" dimension style. As a result, they form a family.

A Sub-style is permanent, unlike the Override command, which is temporary.

Note: This sounds much more complicated than it is. Just follow the steps below. It is very easy.

How to create a sub-style for Radius dimensions.

You will set the center mark to None for the Radius command only.
The Diameter command center mark will not change.

1. Open your **Border A** drawing.

2. Select the **DIMENSION / STYLE** command. (Refer to page 16-8)

3. Select **"Class Style"** from the Style List.

4. Select the **NEW** button.

Continued on the next page...

CREATING A DIMENSION SUB-STYLE....continued

5. Change the "Use for" to: **Radius Dimensions**

6. Select **Continue**

This will turn gray. That's OK

7. Select the **Symbols and Arrows** tab.

8. Change the "Center Marks to **NONE**

9. Select the **OK** button

10. You now have a sub-style that will <u>automatically</u> <u>override</u> the basic "Class Style" whenever you use a Radius command. Diameter dimensions will have centermarks and Radius dimensions will not.

EXERCISE 18A

INSTRUCTIONS:

1. Open **Border A**
2. Draw the objects below using layer Object line.
3. Dimension as shown using Linear, Diameter and Radius. Use Layer Dimension
4. Edit the Title and Ex-XX by double clicking on the text. Do not erase and replace.
5. Save as **EX18A**
6. Plot using Page Setup **Class Model A**

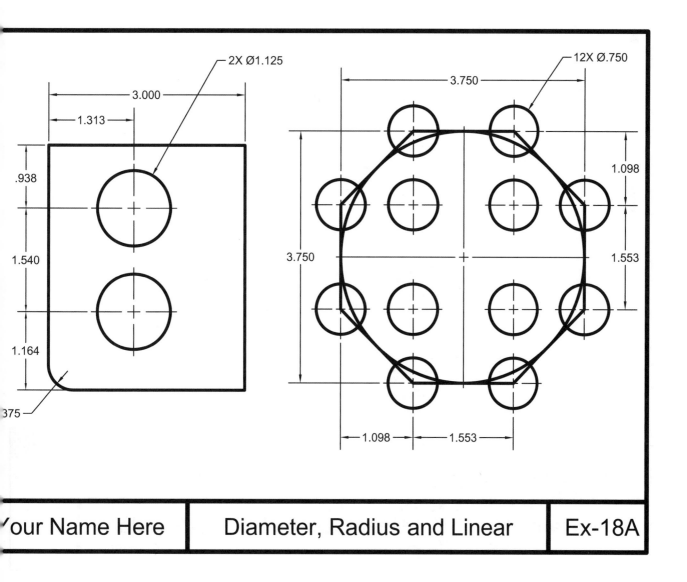

| Your Name Here | Diameter, Radius and Linear | Ex-18A |

EXERCISE 18B

INSTRUCTIONS:
1. Open **Border A**
2. Draw the objects below using layer Object line.

   Note: Refer to the next page for drawing hints.

3. Dimension as shown using Linear, Diameter and Angular. Use Layer Dimension
4. Edit the Title and Ex-XX by double clicking on the text. Do not erase and replace.
5. Save as **EX18B**
6. Plot using Page Setup **Class Model A**

Refer to the next page for drawing hints

Your Name Here	Linear, Diameter and Angular	Ex-18

EXERCISE 18B-Helper

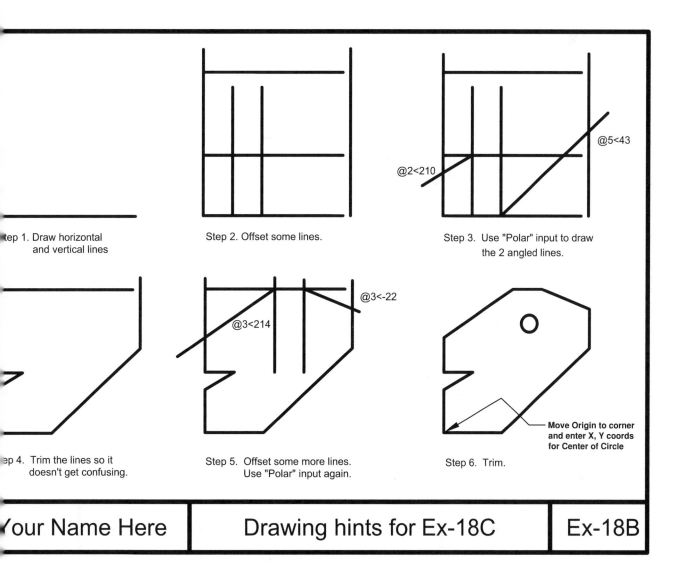

Step 1. Draw horizontal and vertical lines

Step 2. Offset some lines.

Step 3. Use "Polar" input to draw the 2 angled lines.

@5<43

@2<210

Step 4. Trim the lines so it doesn't get confusing.

Step 5. Offset some more lines. Use "Polar" input again.

@3<214

@3<-22

Step 6. Trim.

Move Origin to corner and enter X, Y coords for Center of Circle

| Your Name Here | Drawing hints for Ex-18C | Ex-18B |

EXERCISE 18C

INSTRUCTIONS:
1. Open **Border A**
2. Draw the objects below using layer Object line.
3. Dimension as shown using Linear, Diameter and Radius. Use Layer Dimension
4. Edit the Title and Ex-XX by double clicking on the text. Do not erase and replace.
5. Save as **EX18C**
6. Plot using Page Setup **Class Model A**

Your Name Here	More Practice	Ex-18

LEARNING OBJECTIVES

After completing this lesson, you will be able to:

1. Create Multileaders
2. Modify Multileader lines
3. Create a New Multileader Style
4. Dimension angular surfaces
5. Add Special Characters
6. Pre-assign a Prefix or Suffix to a dimension.

LESSON 19

MULTILEADER

A multileader is a single object consisting of a **pointer**, **line**, **landing** and **content**. You may draw a multileader, Pointer first, tail first or content first.

To select the Multileader use:
 Annotate tab / Leaders panel

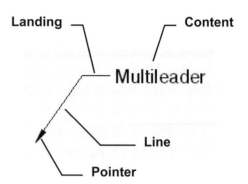

Create a single Multileader

1. Select the **Multileader** tool

    Command: _mleader
2. Specify leader arrowhead location or [leader Landing first/Content first/Options] <Options>: *specify arrowhead location (P1)*
3. Specify leader landing location: *specify the landing location (P2)*
4. The Multitext editor appears. *Enter text then select "Close Text Editor"*

Add leader lines to existing Multileader

1. Select the **Add Leader** tool
2. Select a multileader: *click on the existing multileader line (P1)*
3. Specify leader arrowhead location: *specify arrowhead location (P2)*
4. Specify leader arrowhead location: *specify next arrowhead location (P3)*
5. Specify leader arrowhead location: *specify next arrowhead location (P4)*
6. Specify leader arrowhead location: *press <enter> to stop*

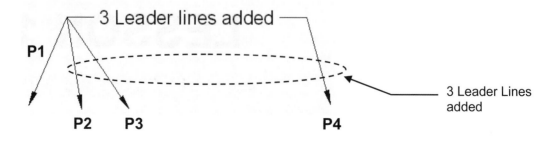

MULTILEADER....continued

Remove a leader line from an existing Multileader

1. Select the **Remove Leader** tool

2. Select a multileader: ***click anywhere on the existing multileader***
3. Specify leaders to remove: ***select the leader line to remove***
4. Specify leaders to remove: ***press <enter> to stop***

Align Multileaders

1. Select the **Align Multileader** tool

Command: _mleaderalign
2. Select multileaders: ***select the leaders to align (Note 1 and Note 3)***
3. Select multileaders: ***press <enter> to stop selecting***
   Current mode: Use current spacing
4. Select multileader to align to or [Options]: ***select the leader to <u>align to</u> (Note 2)***
5. Specify direction: ***move the cursor and press left mouse button***

Before Align   **After Align**

Note: <u>Collect Multileader</u> works best with Blocks and will be
discussed in another lesson.

CREATE A MULTILEADER STYLE

You may create **Multileader styles** to control it's appearance.
This will be similar to creating a dimension style.

1. Open **Border A**

2. Select the **Multileader Style Manager** by selecting the ↘ on the Leader panel.

3. Select the **New** button.
4. Enter the New Style name: **Class ML style**
5. Select the **Continue** button.

6. Select the **Leader Format** tab and change the settings as shown below.

General – You may set the leader
line to straight or spline (curved).
You may change the color, linetype
and lineweight.

Arrowhead – Select the symbol for
the pointer and size of pointer.

Leader break - specify the size of
the gap in the leader line if
Dimension break is used.
(Refer to page 17-7 for Dimension
break reminder)

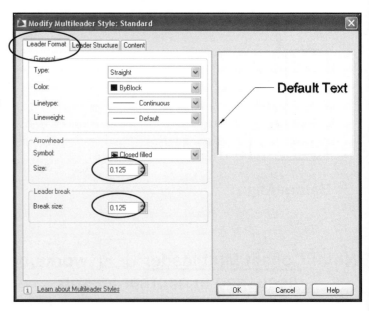

CREATE A MULTILEADER STYLE....continued

7. Select the **Leader Structure** tab and change the settings as shown below.

Constraints – Specify how many line segments to allow and on what angle.

Landing Settings – Specify the length of the Landing (the horizontal line next to the note) or turn it off by unchecking the "Automatically include landing" box.

Scale – "Annotation" will be discussed in Lesson 27.

8. Select the **Content** tab and change the settings as shown below.

Multileader type – Select what to attach to the landing.

Text Options – Specify the leader note appearance.

Leader Connection – Specify how the note attaches to the landing.

9. Select the **OK** button.

The **Class ML style** should be listed in the "Style" area.

10. Select **Set Current** button.

11. Select **Close** button.

12. **Save** your **Border A** again.
Now whenever you use **Border A** the Multileader style **Class ML style** will be there.

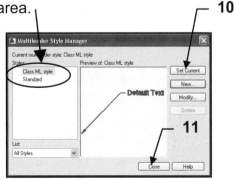

ALIGNED DIMENSIONING

The **ALIGNED** dimension command aligns the dimension with the angle of the object that you are dimensioning. The process is the same as Linear dimensioning. It requires two extension line origins and the placement of the dimension text location. (Example below)

1. Select the **ALIGNED** command using one of the following:

> **Ribbon = Annotate tab / Dimension panel /**
>
> **Keyboard = Dal <enter>**

Command: _dimaligned

2. Specify first extension line origin or <select object>: *select the first extension line origin (P1)*

3. Specify second extension line origin: *select the second extension line origin (P2)*

4. Specify dimension line location or [Mtext/Text/Angle]: *place dimension text location (P3)*

Dimension text = *the dimension value will appear here*

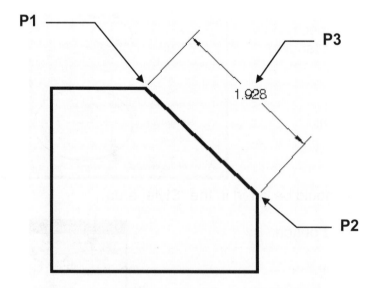

SPECIAL TEXT CHARACTERS

Characters such as the *degree symbol (°), diameter symbol (Ø)* and the *plus / minus symbols (±)* are created by typing **%%** and then the appropriate "**code**" letter.

SYMBOL		CODE
Ø	Diameter	%%C
°	Degree	%%D
±	Plus / Minus	%%P

SINGLE LINE TEXT

When using "Single Line Text", type the code in the sentence. After you enter the code the symbol will appear.

> **For example:**
> Entering <u>350**%%D**</u> will create: **350°**. The **"D"** is the **"code"** letter for degree.

MULTILINE TEXT

When using Multiline Text you may enter the code in the sentence or you may select a symbol using the **Symbol tool** located on the **Insert panel.**

Insert Tool

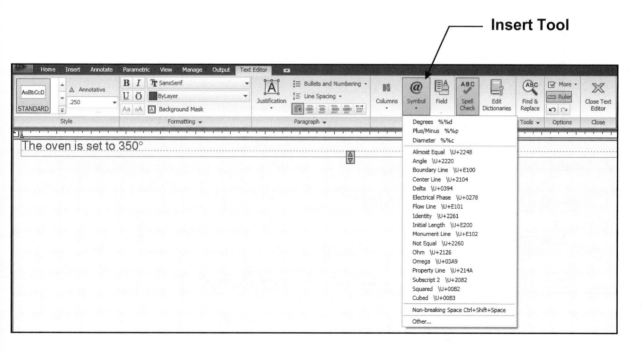

PREFIX and SUFFIX

The **PREFIX** (before) and the **SUFFIX** (after) allows you to preset text to be inserted automatically as you dimension.

Note:
You would use this if you had multiple dimensions that included a prefix or a suffix. If you had only 1 or 2 you would merely edit the dimension text as shown on page 17-2.

1. Select **Dimension Style**.

2. Select the **Primary Units** tab.

3. Enter the **Prefix** or **Suffix** content.

    Example:

No Prefix/Suffix   Prefix=2X Suffix=Ref

Note: If you enter text in the Prefix box when drawing Radial dimensions, the **"R"** for radius or the **diameter symbol**, <u>will not</u> be automatically drawn. You must also add the letter <u>R</u> or symbol code to the prefix box.   Example: **2X R.   Or   2X %%C**

EXERCISE 19A

INSTRUCTIONS:

1. Open **Border A**
2. Draw the Lines as shown using Polar coordinates and Dynamic Input.
3. Use Layer Object lines.
4. Dimension using Linear, Aligned and Angular.
5. Use Dimension Style "Class Style" and Layer Dimension
6. Edit the Title and Ex-XX by double clicking on the text. Do not erase and replace.
7. Save as **EX19A**
8. Plot using Page Setup **Class Model A**

| Your Name Here | Aligned dimensions | Ex-19A |

EXERCISE 19B

INSTRUCTIONS:
1. Open **Border A**
2. Draw the object shown below.
3. Use Layer Object lines.
4. Dimension as shown.
5. Use Dimension Style "Class Style" and Layer Dimension.
6. For the Note: "Special Instructions to Baker" use: Text Ht. = .125 Layer = Text
7. Edit the Title and Ex-XX by double clicking on the text. Do not erase and replace.
8. Save as **EX19B**
9. Plot using Page Setup **Class Model A**

Special Instructions to Baker
1. Ingredients:
   Sugar, Biquick and chocolate chips.
2. Bake at 350° for 30 minutes.
3. Cool 10 minutes.
4. Eat (Only if you have been good).

Note: Break Sharp Edge

7.631

70° Ref

2.610

.750

20° Ref

4.500

7.171

Ø .750 Chocolate chip

Note: This is a dimension not a Leader. Refer to page 17-2 to edit.

| Your Name Here | Triangular Cookie | Ex-19 |

EXERCISE 19C

INSTRUCTIONS:

1. Open **Border A**
2. Create the Multileader style shown on page 19-4 if you haven't already.
3. Draw the object shown below.
4. Use Layer Object lines.
5. Dimension as shown.
6. Use Dimension Style "Class Style" and "Class ML style" and Layer Dimension.
7. Edit the Title and Ex-XX by double clicking on the text. Do not erase and replace.
8. Save as **EX19C**
9. Plot using Page Setup **Class Model A**

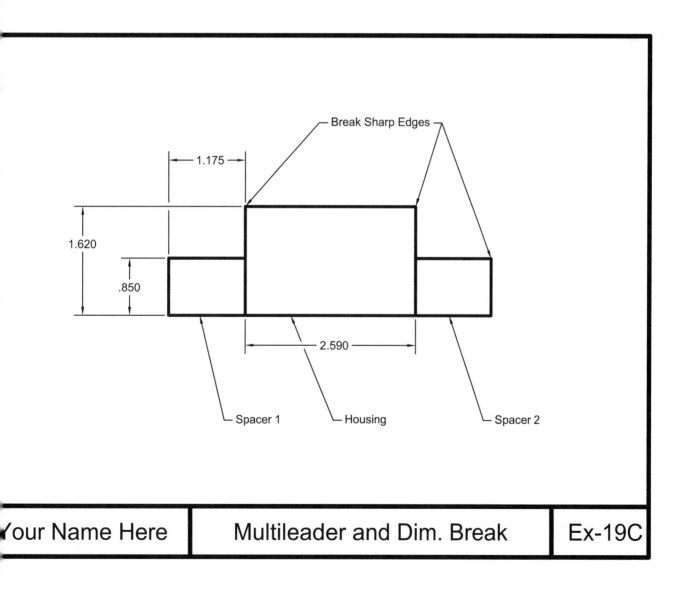

| Your Name Here | Multileader and Dim. Break | Ex-19C |

Notes:

LEARNING OBJECTIVES

After completing this lesson, you will be able to:

1. Use Quick Dimension to automatically dimension
2. Edit Multiple Dimensions
3. Edit extension locations using grips

LESSON 20

QUICK DIMENSION

Quick Dimension creates multiple dimensions with one command. Quick Dimension can create Continuous, Staggered, Baseline, Ordinate, Radius and Diameter dimensions. Qdim only works in modelspace. It is not trans-spatial. (Refer to Lesson 27)

The following is an example of how to use the "Continuous" option. Staggered, Baseline, Diameter and Radius are explained on the following pages.

1. Select the **Quick Dimension** command using one of the following:

   Ribbon = Annotate tab / Dimension panel / [icon]

   Keyboard = Qdim <enter>

   Command: _qdim
   Associative dimension priority = Endpoint
   Select geometry to dimension

2. Select the objects to be dimensioned with a crossing window or pick each object

2. Crossing Window

3. Press <enter> to stop selecting objects.

4. Specify dimension line position, or
   [Continuous/Staggered/Baseline/Ordinate/Radius/Diameter/datumPoint/Edit/settings]<Continuous>: *Select "C" <enter> for Continuous.*

4. Specify dimension line location.

| ⊢—1.350—⊣ | ⊢—1.422—⊣ | ⊢—1.133—⊣ |

QUICK DIMENSION....continued

STAGGERED

BASELINE

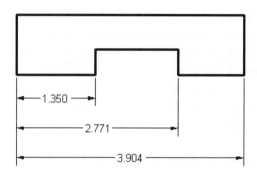

DIAMETER

1. You can use a crossing window. Qdim will automatically filter out any linear dims.
2. <u>Dimension Line Length</u> is determined by the "Baseline Spacing" setting in the Dimension Style.

RADIUS

1. You can use a crossing window. Qdim will automatically filter out any linear dims.
2. Dimension Line Length is determined by the "Baseline Spacing" setting in the Dimension Style.

EDITING MULTIPLE DIMENSIONS

You can edit existing multiple dimensions using the QDIM / Edit command. The Qdim, edit command will edit all multiple dimensions, no matter whether they were created originally with Qdim or not. All multiple: linear, baseline and continue dimensions respond to this editing command.

The example below illustrates changing Baseline dimensions to Continuous.

1. Select the **Quick Dimension** command

2. Select the dimensions to edit.

2. Select dimensions to edit.

3. Press <enter> to stop selecting dimensions.

4. Select an option, such as Continuous.

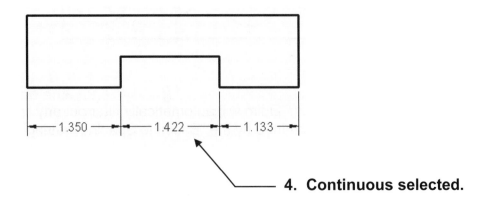

4. Continuous selected.

EDITING DIMENSIONS USING GRIPS

Sometimes, when using Quick Dimension, the extension lines point to a location that you do not prefer. If this happens select the dimension, select the grip and drag it to the preferred location. If the dimension is associative (refer to page 16-2) the dimension text will change to reflect the new extension line location.

EXAMPLE:

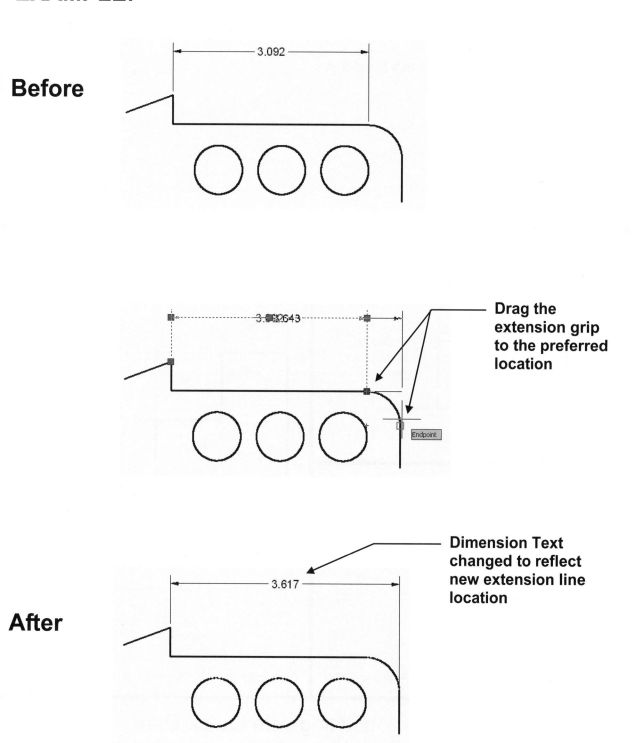

Before

Drag the
extension grip
to the preferred
location

Dimension Text
changed to reflect
new extension line
location

After

EXERCISE 20A

INSTRUCTIONS:

1. Open **EX16A**
2. Edit the dimensions as shown below using Quick Dimension. (Refer to page 20-4)
   Notice: some of the Continuous have changed to Baseline and vice versa. This is a very helpful command.
3. Edit the Title and Ex-XX by double clicking on the text. Do not erase and replace.
4. Save as **EX20A**
5. Plot using Page Setup **Class Model A**

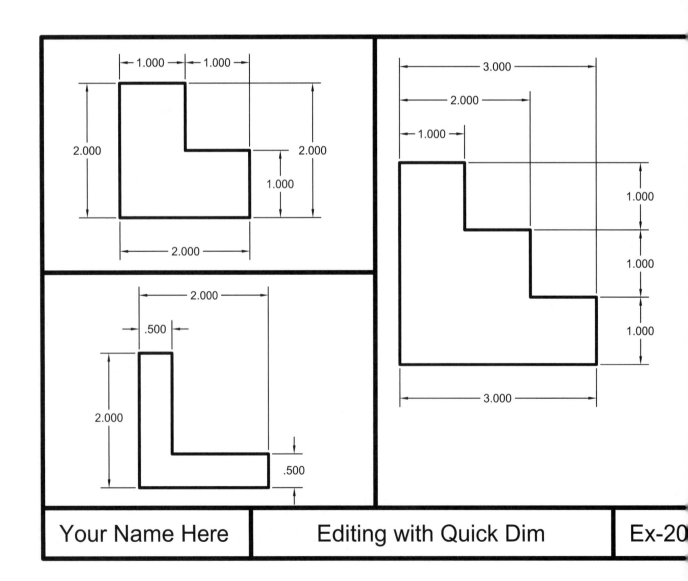

| Your Name Here | Editing with Quick Dim | Ex-20 |

EXERCISE 20B

INSTRUCTIONS:
1. Open **Border A.**
2. Draw the 2 Rectangular forms below.
3. Dimension them using Quick Dimension, Diameter and Radius and a few Linear.

   This time you will have to use "Pick" instead of "Crossing Window" to select the objects to dimension.
   You may also have to move some of the dimension using grips because they may not be exactly where you want them.
   Try it, I think you will find it very interesting.

4. Edit the Title and Ex-XX by double clicking on the text. Do not erase and replace.
5. Save as **EX20B**
6. Plot using Page Setup **Class Model A**

| Your Name Here | Dimensioning with Quick Dim | Ex-20B |

EXERCISE 20C

INSTRUCTIONS:

1. Open **Border A**
2. Draw the object as shown below using Layer Object line.
3. Dimension using Quick Dimension whenever possible.
   You may have to use "grips" to move the extension lines to the correct endpoint on the object. And you may have to mix in a few Linear dimensions. Experiment and try some new methods.
4. Use Dimension Style "Class Style" and Layer Dimension
5. Edit the Title and Ex-XX by double clicking on the text. Do not erase and replace.
6. Save as **EX20C**
7. Plot using Page Setup **Class Model A**

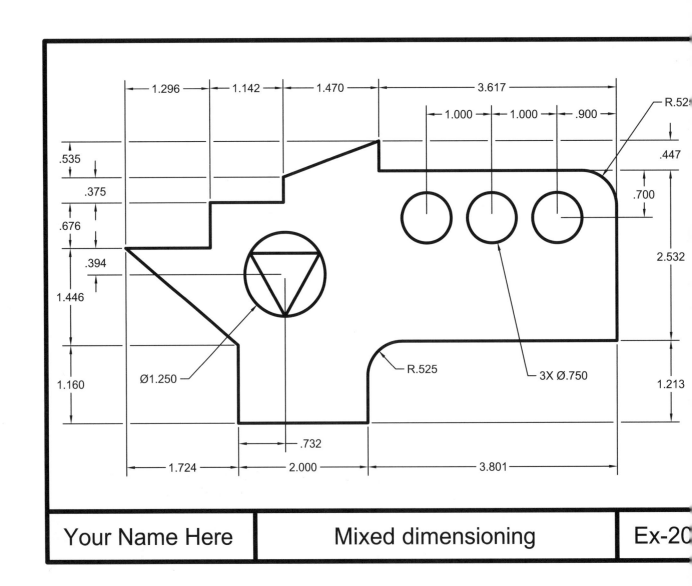

| Your Name Here | Mixed dimensioning | Ex-20 |

LEARNING OBJECTIVES

After completing this lesson, you will be able to:

1. Match an objects properties with another object
2. Create a Revision Cloud
3. Select a Revision Cloud Style
4. "Wipe Out" an area of an object

LESSON 21

MATCH PROPERTIES

Match Properties is used to "paint" the properties of one object to another. This is a simple and useful command. You first select the object that has the desired properties (the source object) and then select the object you want to "paint" the properties to (destination object).

Only one "source object" can be selected but its properties can be painted to any number of "destination objects".

1. Select the Match Properties command using one of the following:

   Ribbon = Home tab / Clipboard panel /

   Keyboard = MA <enter>

   Command: matchprop
2. Select source object: ***select the object with the desired properties to match***

3. Select destination object(s) or [Settings]: ***select the object(s) you want to receive the matching properties.***

4. Select destination object(s) or [Settings]: ***select more objects or <enter> to stop.***

Note: If you do not want to match all of the properties, after you have selected the source object, right click and select "Settings" from the short cut menu, before selecting the destination object. Uncheck all the properties you do not want to match and select the OK button. Then select the destination object(s).

LAYER MATCH

If you draw an object on the wrong layer you can easily change it to the desired layer using Layer Match command. You first select the object that needs to be changed and then select an object that is on the correct layer (object on destination layer).

1. Select the Layer Match command using one of the following:

    Ribbon = Home tab / Layers panel /

    Browser Menu = Format / Layer Tools / Layer Match

    Keyboard = Laymch <enter>

    Command: _laymch
2. Select objects to be changed:*select the objects that need to be changed*

3. Select objects: *select more objects or <enter> to stop selecting*

4. Select object on destination layer or [Name]: *select the object that is on the layer that you want to change to*

Note:
You may also easily change the layer of an object to another layer as follows:

1. Select the object that you wish to change.

2. Select the Layer drop down menu and select the layer that you wish the object to be placed on.

For example:
If you had a dimension on the object line layer by mistake.

1. Select the dimension that is mistakenly on the object line layer.

2. Select layer dimension from the drop down menu.

CREATING A REVISION CLOUD

When you make a revision to a drawing it is sometimes helpful to highlight the revision for someone viewing the drawing. A common method to highlight the area is to draw a "Revision Cloud" around the revised area. This can be accomplished easily with the "Revision Cloud" command.

The Revision Cloud command creates a series of sequential arcs to form a cloud shaped object. You set the minimum and maximum arc lengths. (Maximum arc length cannot exceed three times the minimum arc length. Example: Min = 1, Max can be 3 or less) If you set the minimum and maximum different lengths the arcs will vary in size and will display an irregular appearance.

Min & Max same length   **Min & Max different length**

To draw a Revision Cloud you specify the start point with a left click then drag the cursor to form the outline. AutoCAD automatically draws the arcs. When the cursor gets very close to the start point, AutoCAD snaps the last arc to the first arc and closes the shape.

1. Select the Revision Cloud command using one of the following:

    Ribbon = Home tab / Draw ▼ panel /

    Keyboard = Revcloud <enter>

    Command: _revcloud
    Minimum arc length: .50   Maximum arc length: .50   Style: Normal

2. Specify start point or [Arc length/Object/Style] <Object>: *Select "Arc length"*

3. Specify minimum length of arc <.50>: *Specify the minimum arc length*

4. Specify maximum length of arc <.50>: *Specify the maximum arc length*

5. Specify start point or [Arc length/Object/Style] <Object>: *Place cursor at start location & left click.*

6. Guide crosshairs along cloud path...*Move the cursor to create the cloud outline.*

7. Revision cloud finished. *When the cursor approaches the start point, the cloud closes automatically.*

CONVERT A CLOSED OBJECT TO A REVISION CLOUD

You can convert a closed object, such as a circle, ellipse, rectangle or closed polyline to a revision cloud. The original object is deleted when it is converted.

(If you want the original object to remain, in addition to the new rev cloud, set the variable "delobj" to "0". The default setting is "1".)

1. Draw a closed object such as a circle.

2. Select the Revision Cloud command using one of the following:

    Ribbon = Home tab / Draw ▾ panel /

    Keyboard = Revcloud <enter>

    Command: _revcloud
    Minimum arc length: .50   Maximum arc length: .50   Style: Normal

3. Specify start point or [Arc length/Object/Style] <Object>: ***Select "Arc length"***

4. Specify minimum length of arc <.50>: ***Specify the minimum arc length***

5. Specify maximum length of arc <.50>: ***Specify the maximum arc length***

6. Specify start point or [Arc length/Object/Style] <Object>: ***Select "Object".***

7. Select object: ***Select the object to convert***

8. Select object: Reverse direction [Yes/No] <No>: ***Select Yes or No***
    Revision cloud finished.

Reverse direction?

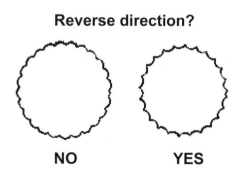

NO   YES

NOTE:
The Match Properties command will not match the arc length from the source cloud to the destination cloud.

REVISION CLOUD STYLE

You may select one of 2 styles for the Revision Cloud; **Normal** or **Calligraphy**.
Normal will draw the cloud with one line width.
Calligraphy will draw the cloud with variable line widths to appear as though you used a chiseled calligraphy pen.

Normal   **Calligraphy**

1. Select the Revision Cloud command using one of the following:

   Ribbon = Home tab / Draw ▼ panel /

   Keyboard = Revcloud <enter>

   Command: _revcloud
   Minimum arc length: .50   Maximum arc length: 1.00   Style: Normal

2. Specify start point or [Arc length/Object/Style] <Object>: *Select "Style"<enter>*

3. Select arc style [Normal/Calligraphy] <Calligraphy>:*Select "N or C"<enter>*

4. Specify start point or [Arc length/Object/Style] <Object>: *Select "Arc length"*

5. Specify minimum length of arc <.50>: *Specify the minimum arc length*

6. Specify maximum length of arc <1.00>: *Specify the maximum arc length*

7. Specify start point or [Object] <Object>: *Place cursor at start location & left click.*

8. Guide crosshairs along cloud path...*Move the cursor to create the cloud outline.*

9. Revision cloud finished. *When the cursor approaches the start point, the cloud closes automatically.*

WIPEOUT

The Wipeout command creates a blank area that covers existing objects. The area has a background that matches the background of the drawing area. This area is bounded by the wipeout frame, which you can turn on or off.

1. Select the Wipeout command using one of the following:

   Ribbon = Home tab / Draw ▾ panel / 🗔

   Keyboard = Wipeout <enter>

2. Command: _wipeout Specify first point or [Frames/Polyline] <Polyline>:
   specify the first point of the shape (P1)
3. Specify next point: ***specify the next point (P2)***
4. Specify next point or [Undo]: ***specify the next point (P3)***
5. Specify next point or [Undo]: ***specify the next point or <enter> to stop***

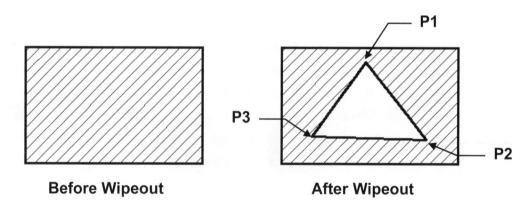

Before Wipeout   **After Wipeout**

TURNING FRAMES ON OR OFF

1. Select the Wipeout command.
2. Select the "Frames" option.
3. Enter ON or OFF.

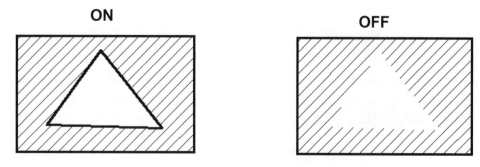

ON   OFF

Note: If you want to move the objects and the wipeout area, you must select both and move them at the same time. Do not move them separately.

EXERCISE 21A

INSTRUCTIONS:
1. Open **EX5F.dwg**
2. Using **Match Properties,** change the properties of the Polygon to the same properties as the Donut.
3. Using **Wipeout**, block out the bottom donut approximately as shown.
4. Edit the Title and Ex-XX by double clicking on the text. Do not erase and replace.
5. Save as **EX21A**
6. Plot using Page Setup **Class Model A**

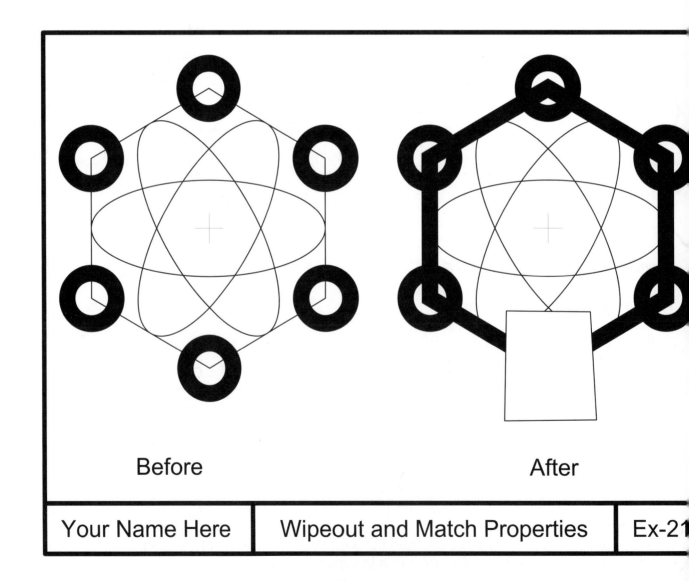

Before

After

| Your Name Here | Wipeout and Match Properties | Ex-21 |

EXERCISE 21B

INSTRUCTIONS:

1. Open **Border A**
2. Draw the Revision clouds below:
   - #1: Set Min. and Max. Arc Length to .500
   - #2: Set Min. Arc Length to .500 and Max. Arc Length to 1.00
   - #3: Same as #2 but change Style to Calligraphy.
3. Draw (2) 2.50 Diameter Circles and Convert to Revision clouds.
   - Set Min and Max Arc Length to .50   Set Style to Normal
   - #4 Reverse = No   #5 Reverse = Yes
4. Use Layer Object line.
5. Edit the Title and Ex-XX by double clicking on the text. Do not erase and replace.
6. Save as **EX21B**
7. Plot using Page Setup **Class Model A**

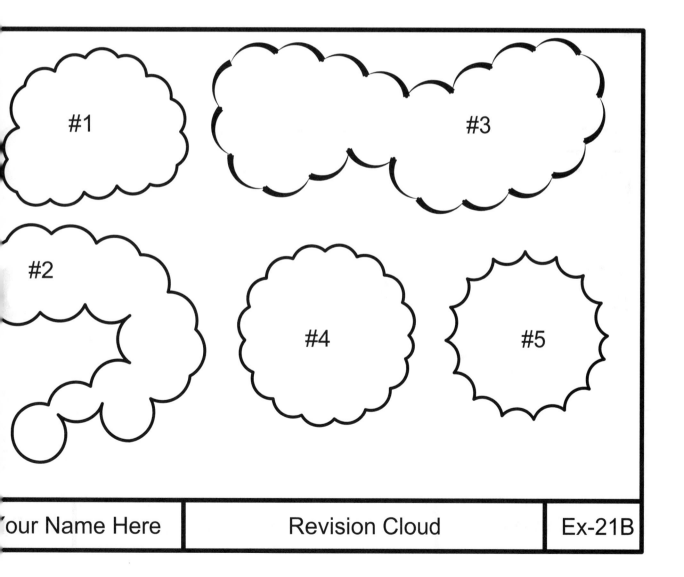

| Your Name Here | Revision Cloud | Ex-21B |

21-9

EXERCISE 21C

INSTRUCTIONS:

1. Open **Border A**
2. The drawing below is designed to give you more practice drawing objects and dimensioning. Use any method that you have learned in the previous lessons.
3. Edit the Title and Ex-XX by double clicking on the text. Do not erase and replace.
4. Save as **EX21C**
5. Plot using Page Setup **Class Model A**

| Your Name Here | Base Plate | Ex-21 |

EXERCISE 21D

INSTRUCTIONS:

1. Open **Border A**
2. The drawing below is designed to give you more practice drawing objects and dimensioning. Use any method that you have learned in the previous lessons.
3. Edit the Title and Ex-XX by double clicking on the text. Do not erase and replace.
4. Save as **EX21D**
5. Plot using Page Setup **Class Model A**

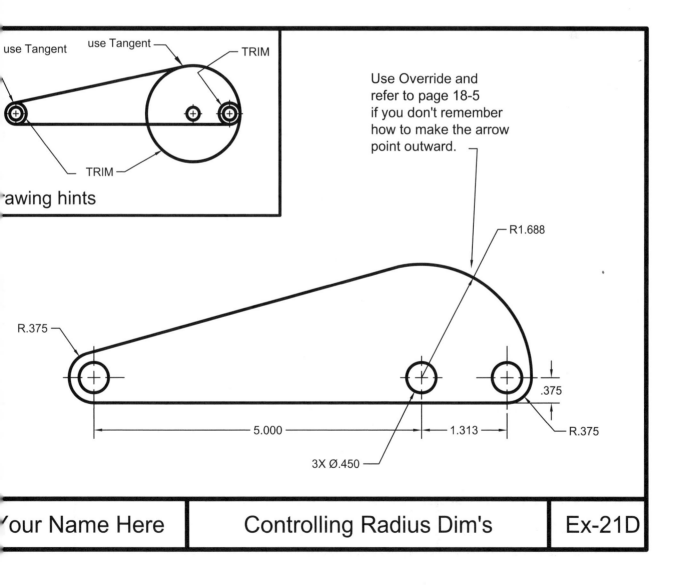

EXERCISE 21E

INSTRUCTIONS:
1. Open **Border A**
2. The drawing below is designed to give you more practice drawing objects and dimensioning. Use any method that you have learned in the previous lessons.
3. Edit the Title and Ex-XX by double clicking on the text. Do not erase and replace.
4. Save as **EX21E**
5. Plot using Page Setup **Class Model A**

Draw 1 segment then use polar array.

Draw 1 slot then use polar array then trim.

Construction Hints

R1.313

.300
.150

Ø4.875

1.110

.600 .300

R2.063

R.938

Use Override
See page 17-5 and 18-5
or use Flip Arrow

Use "Flip Arrow"
see page 18-8

R.188

| Your Name Here | Circular Pattern | Ex-21 |

LEARNING OBJECTIVES

After completing this lesson, you will be able to:

1. Draw an Arc using 10 different methods
2. Dimension a Arc segments

LESSON 22

ARC

There are 10 ways to draw an ARC in AutoCAD.   Not all of the ARCS options are easy to create so you may find it is often easier to **trim a Circle** or use the **Fillet** command.

On the job, you will probably only use 2 of these methods. Which 2 depends on the application.

An **ARC** is a <u>segment</u> of a circle and <u>must be less than 360 degrees</u>.

Most ARCS are drawn counter-clockwise but some may be drawn clockwise by entering a negative input.

1. Select the Arc Command using one of the following:

       Ribbon = Home tab / Draw panel /

       Keyboard = A <enter>

2. Refer to Exercises 22A through 22J for examples of each method listed below.

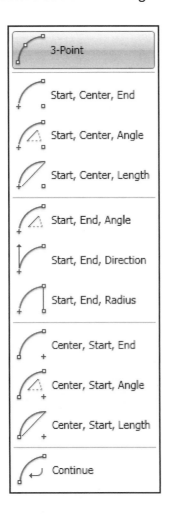

DIMENSIONING ARC LENGTHS

You may dimension the distance along an Arc. This is known as the **Arc length**.
Arc length is an <u>associative</u> dimension.

Example:

1. Select the Arc length dimension command using one of the following:

Ribbon = Annotate tab / Dimension panel /   **Select ▼**

Keyboard = dimarc <enter>

    Command: _dimarc
2. Select arc or polyline arc segment: *select the Arc*
3. Specify arc length dimension location, or [Mtext/Text/Angle/Partial/Leader]:
        place the dimension line and text location

    Dimension text = *dimension value will be shown here*

To differentiate the Arc length dimensions from Linear or Angular dimensions, arc length
dimensions display an arc (⌒) symbol by default. (Also called a "hat" or "cap")

The arc symbol may be displayed either <u>above,</u> or <u>preceding</u> the dimension text.
You may also choose not to display the arc symbol.

Example:

     PRECEDING   **ABOVE**   **NONE**

Specify the placement of the arc symbol in the **Dimension Style / Symbols and Arrows**
tab or you may edit its position using the Properties Palette.

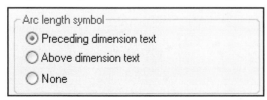

Continued on the next page...

DIMENSIONING ARC LENGTHS....continued

The extension lines of an Arc length dimension are displayed as <u>radial</u> if the included angle is <u>greater than 90 degrees</u>.

Example:

The extension lines of an Arc length dimension are displayed as <u>orthogonal</u> if the included angle is <u>less than 90 degrees</u>.

Example:

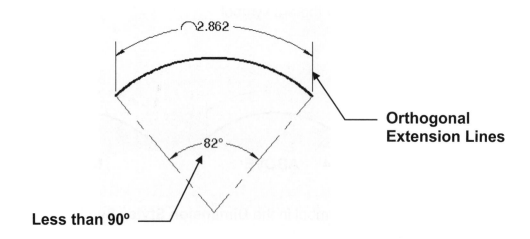

DIMENSIONING A LARGE CURVE

When dimensioning an arc the dimension line should pass through the center of the arc. However, for large curves, the true center of the arc could be very far away, even off the sheet.

When the true center location cannot be displayed you can create a "Jogged" radius dimension.

EXAMPLE:

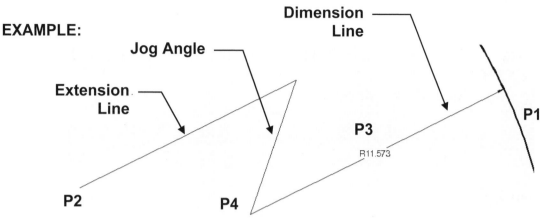

You can specify the jog angle in the Dimension Style / Symbols and Arrows tab.

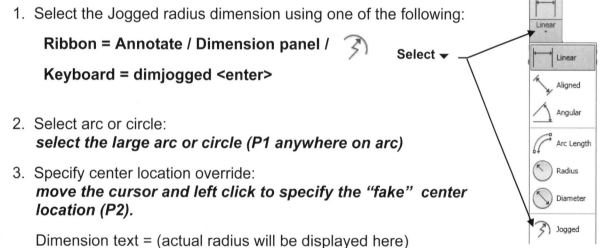

1. Select the Jogged radius dimension using one of the following:

   Ribbon = Annotate / Dimension panel /

   Keyboard = dimjogged <enter>

2. Select arc or circle:
   select the large arc or circle (P1 anywhere on arc)

3. Specify center location override:
   move the cursor and left click to specify the "fake" center location (P2).

   Dimension text = (actual radius will be displayed here)

4. Specify dimension line location or [Mtext/Text/Angle]: ***move the cursor and left click to specify the location for the dimension text (P3).***

5. Specify jog location: ***move the cursor and left click to specify the location for the jog (P4).***

CONTROLLING THE JOG

You can set the Jog Angle and Height factor in:
Dimension Style / Symbols and Arrows tab

EXERCISE 22A

INSTRUCTIONS:

1. Open **Border A**
2. Draw the center lines first on layer Center line.
3. Draw the **3 Point** Arcs using Layer Object Line.
   Place the **1st point** at location ① **2nd point** at location ② and **3rd point** at location
4. Dimension as shown using Dimension Style Class Style and Layer Dimension.
5. Edit the Title and Ex-XX by double clicking on the text. Do not erase and replace.
6. Save as **EX22A**
7. Plot using Page Setup **Class Model A**

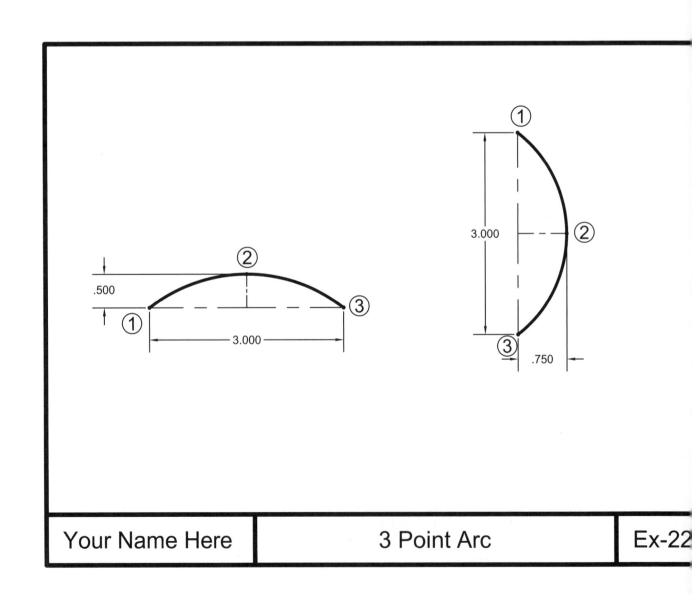

| Your Name Here | 3 Point Arc | Ex-22 |

EXERCISE 22B

INSTRUCTIONS:

1. Open **Border A**
2. Draw the center lines first on layer Center line.
3. Draw the 2 Arcs using method **Start, Center, End** on Layer Object Line.
   Place the **Start** at location ①, **Center** at location ② and **End** at location ③.
4. Dimension as shown using Dimension Style Class Style and Layer Dimension.
5. Edit the Title and Ex-XX by double clicking on the text. Do not erase and replace.
6. Save as **EX22B**
7. Plot using Page Setup **Class Model A**

Note: These arcs are always drawn counter clockwise from Start point.

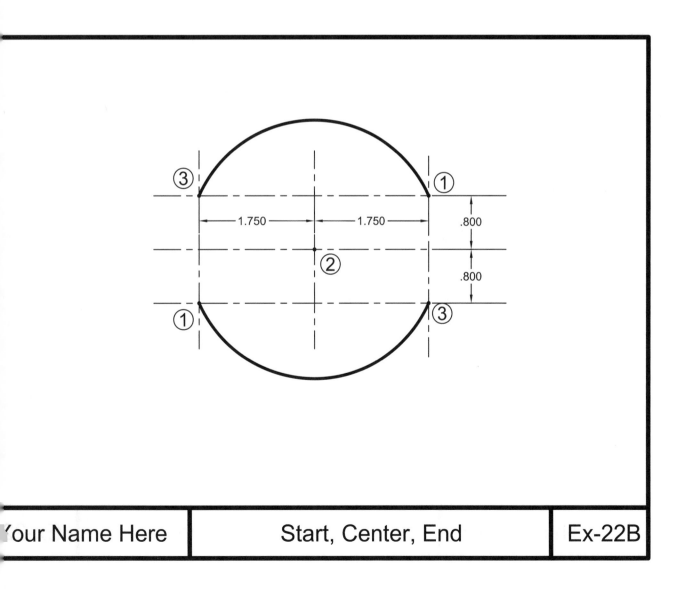

| Your Name Here | Start, Center, End | Ex-22B |

EXERCISE 22C

INSTRUCTIONS:

1. Open **Border A**
2. Draw the center lines first on layer Center line.
3. Draw the 2 Arcs using method **Start, Center, Angle** on Layer Object Line.
   Place the **Start** at location ①, **Center** at location ② and enter **Angle** ③.
4. Dimension as shown using Dimension Style Class Style and Layer Dimension.
5. Edit the Title and Ex-XX by double clicking on the text. Do not erase and replace.
6. Save as **EX22C**
7. Plot using Page Setup **Class Model A**

Note: Positive angles are drawn Counter Clockwise.
Negative angles are drawn Clockwise.

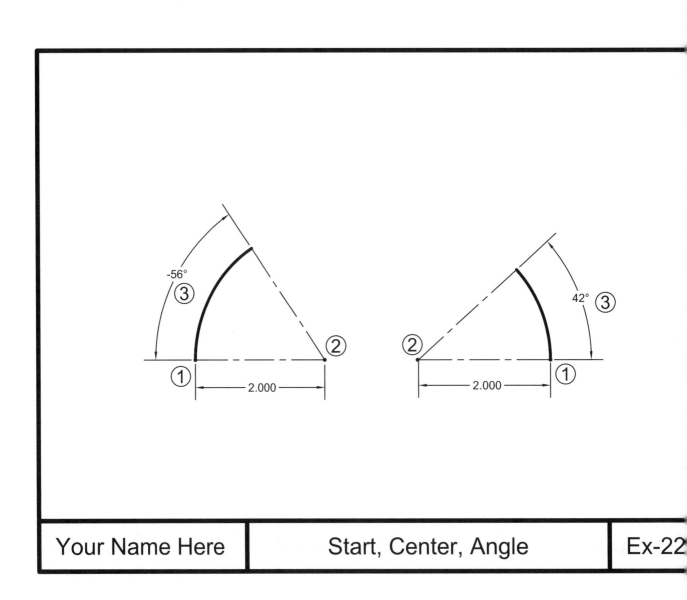

| Your Name Here | Start, Center, Angle | Ex-22 |

EXERCISE 22D

INSTRUCTIONS:

1. Open **Border A**
2. Draw the center lines first on layer Center line.
3. Draw the 2 Arcs using method **Start, Center, Length** on Layer Object Line.
   Place the **Start** at location ①, **Center** at location ② and enter **Length** of Chord ③.
4. Dimension as shown using Dimension Style Class Style and Layer Dimension.
5. Edit the Title and Ex-XX by double clicking on the text. Do not erase and replace.
6. Save as **EX22D**
7. Plot using Page Setup **Class Model A**

Note: Positive Chord length draws the small segment counter clockwise.
Negative Chord length draws the large segment counter clockwise.

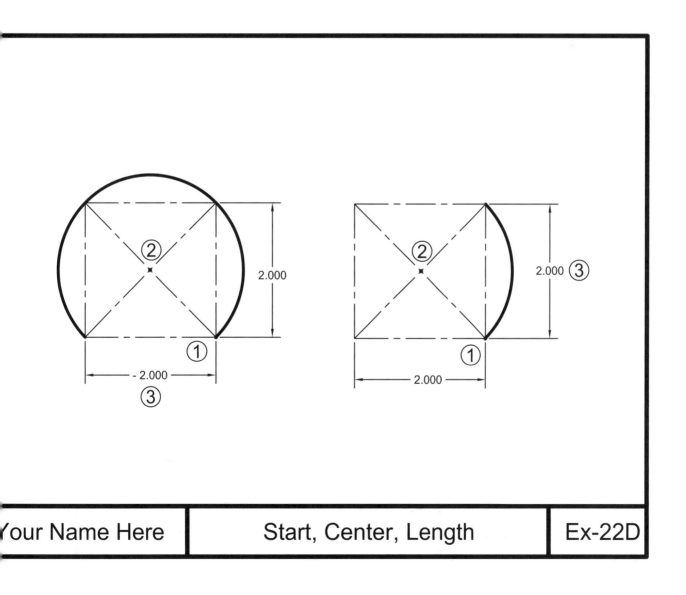

| Your Name Here | Start, Center, Length | Ex-22D |

EXERCISE 22E

INSTRUCTIONS:
1. Open **Border A**
2. Draw the center lines first on layer Center line.
3. Draw the 2 Arcs using method **Start, End, Angle** on Layer Object Line.
   Place the **Start** at location ①, **End** at location ② and enter **Angle** ③.
4. Dimension as shown using Dimension Style Class Style and Layer Dimension.
5. Edit the Title and Ex-XX by double clicking on the text. Do not erase and replace.
6. Save as **EX22E**
7. Plot using Page Setup **Class Model A**

*Note: Positive Angle draws the arc counter clockwise.
Negative Angle draws the arc clockwise.*

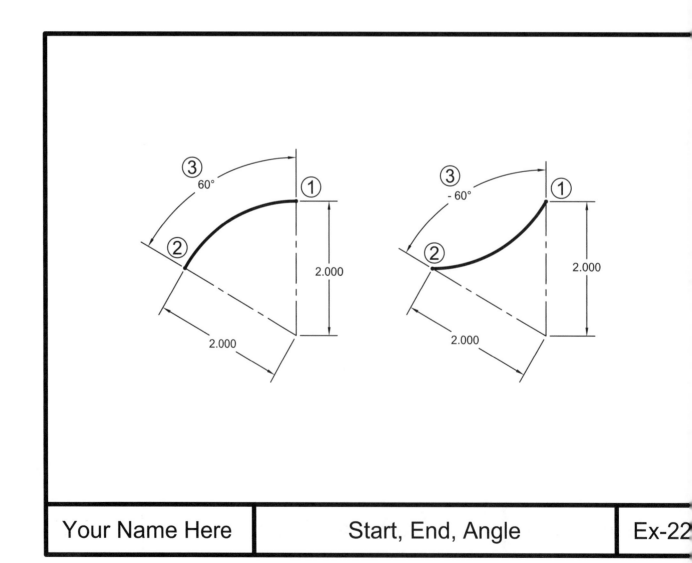

| Your Name Here | Start, End, Angle | Ex-22 |

22-10

EXERCISE 22F

INSTRUCTIONS:

1. Open **Border A**
2. Draw the center lines first on layer Center line.
3. Draw the 2 Arcs using method **Start, End, Direction** on Layer Object Line.
   Place the **Start** at location ①, **End** at location ② and move your cursor in the
   Direction of ③.
4. Dimension as shown using Dimension Style Class Style and Layer Dimension.
5. Edit the Title and Ex-XX by double clicking on the text. Do not erase and replace.
6. Save as **EX22F**
7. Plot using Page Setup **Class Model A**

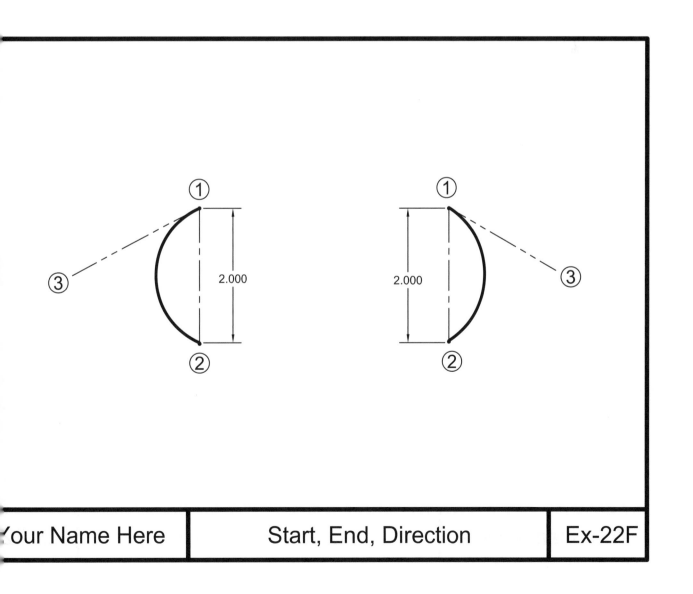

| Your Name Here | Start, End, Direction | Ex-22F |

EXERCISE 22G

INSTRUCTIONS:

1. Open **Border A**
2. Draw the center lines first on layer Center line.
3. Draw the 2 Arcs using method **Start, End, Radius** on Layer Object Line.
   Place the **Start** at location ①, **End** at location ② and enter the **Radius** of ③.
4. Dimension as shown using Dimension Style Class Style and Layer Dimension.
5. Edit the Title and Ex-XX by double clicking on the text. Do not erase and replace.
6. Save as **EX22G**
7. Plot using Page Setup **Class Model A**

   Note: **Positive Radius draws the small segment.**
   Negative Radius draws the large segment.

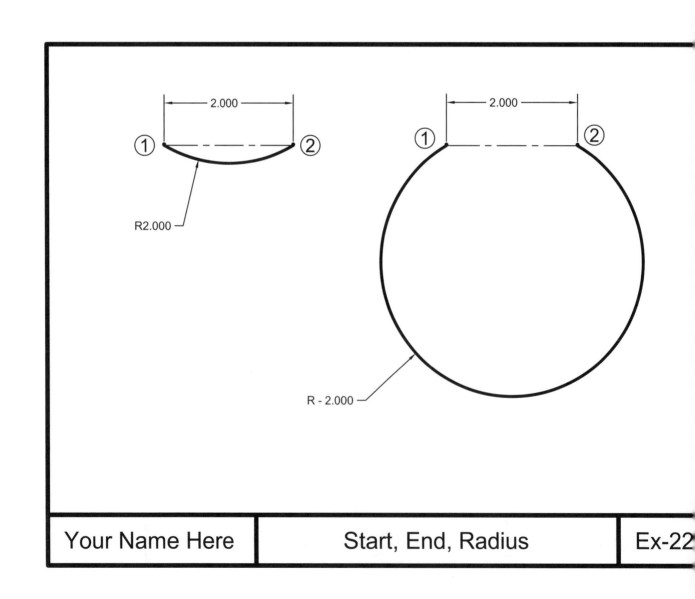

| Your Name Here | Start, End, Radius | Ex-22 |

EXERCISE 22H

INSTRUCTIONS:

1. Open **Border A**
2. Draw the center lines first on layer Center line.
3. Draw the 2 Arcs using method **Center, Start, End** on Layer Object Line.
   Place the **Center** at location ①, **Start** at location ② and **End** at location ③.
4. Dimension as shown using Dimension Style Class Style and Layer Dimension.
5. Edit the Title and Ex-XX by double clicking on the text. Do not erase and replace.
6. Save as **EX22H**
7. Plot using Page Setup **Class Model A**

   Note: Draws the Arc Counter Clockwise only.

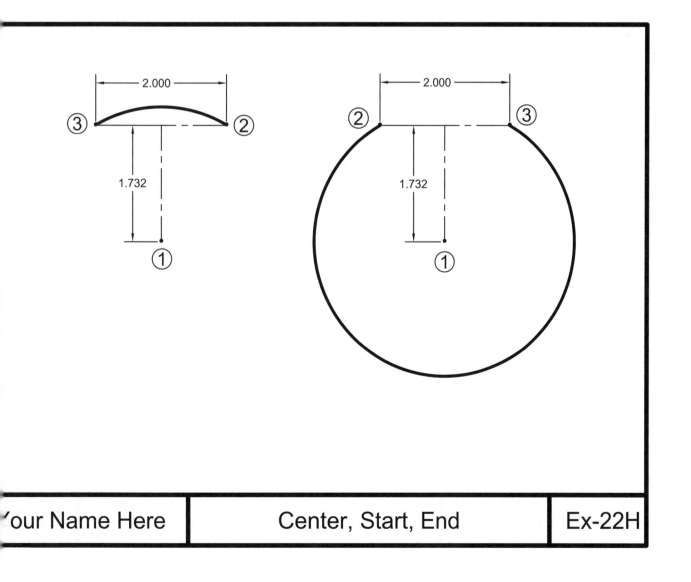

'our Name Here	Center, Start, End	Ex-22H

EXERCISE 22I

INSTRUCTIONS:
1. Open **Border A**
2. Draw the center lines first on layer Center line.
3. Draw the 2 Arcs using method **Center, Start, Angle** on Layer Object Line.
   Place the **Center** at location ① , **Start** at location ② and enter **Angle** ③.
4. Dimension as shown using Dimension Style Class Style and Layer Dimension.
5. Edit the Title and Ex-XX by double clicking on the text. Do not erase and replace.
6. Save as **EX22I**
7. Plot using Page Setup **Class Model A**

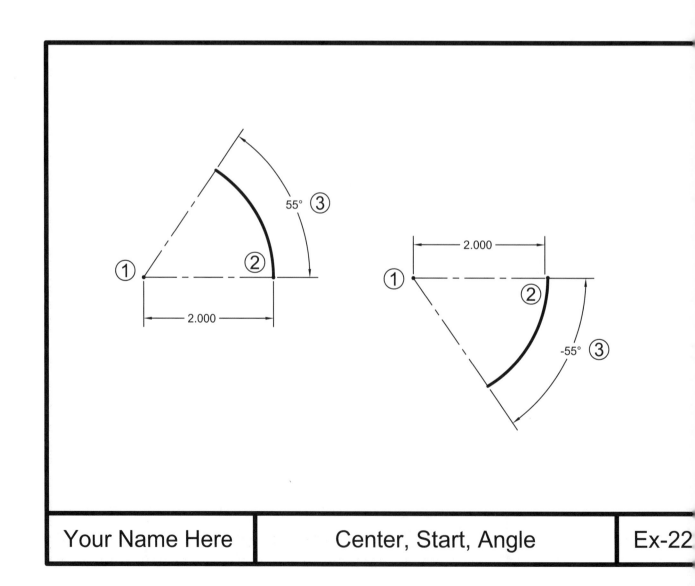

Your Name Here	Center, Start, Angle	Ex-22

EXERCISE 22J

INSTRUCTIONS:
1. Open **Border A**
2. Draw the center lines first on layer Center line.
3. Draw the 2 Arcs using method **Center, Start, Length** on Layer Object Line.
   Place the **Center** at location ①, **Start** at location ② and enter **Length** ③.
4. Dimension as shown using Dimension Style Class Style and Layer Dimension.
5. Edit the Title and Ex-XX by double clicking on the text. Do not erase and replace.
6. Save as **EX22J**
7. Plot using Page Setup **Class Model A**

Note: Positive Chord Length draws small segment counter clockwise.
Negative Chord Length draws large segment counter clockwise.

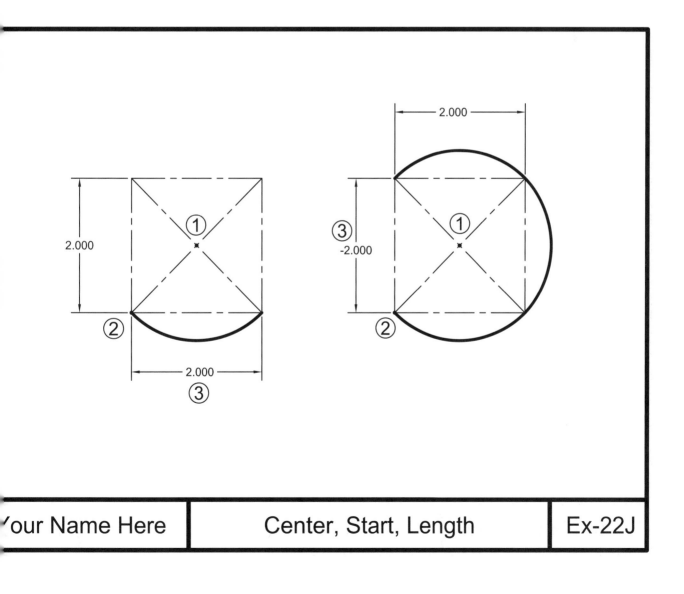

Your Name Here	Center, Start, Length	Ex-22J

EXERCISE 22K

INSTRUCTIONS:
1. Open **Border A**
2. Draw the Lines first on layer Center line.
3. Draw the Arc using Layer Object Line. Refer to the previous pages to select method
4. Dimension as shown using Dimension Style **Class Style** and Layer Dimension.
5. Edit the Title and Ex-XX by double clicking on the text. Do not erase and replace.
6. Save as **EX22K**
7. Plot using Page Setup **Class Model A**

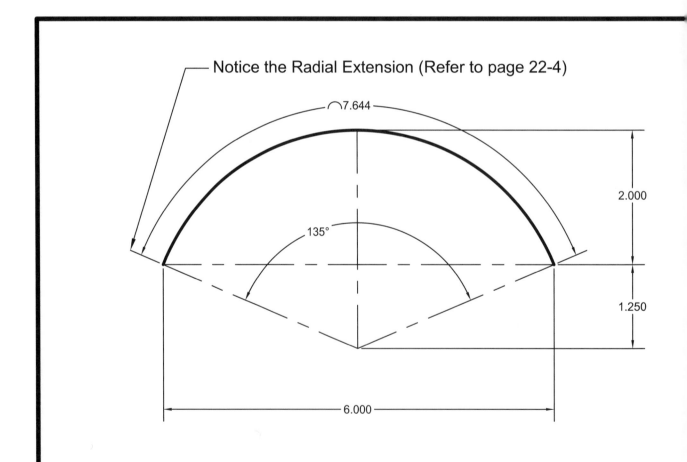

| Your Name Here | Arc Length with Radial extensions | Ex-22 |

EXERCISE 22L

INSTRUCTIONS:

1. Open **Border A**
2. Draw the Lines first on layer Center line.
3. Draw the Arc using Layer Object Line. Refer to the previous pages to select method.
4. Dimension as shown using Dimension Style **Class Style** and Layer Dimension.
5. Edit the Title and Ex-XX by double clicking on the text. Do not erase and replace.
6. Save as **EX22L**
7. Plot using Page Setup **Class Model A**

Notice the Orthogonal Extensions (Refer to page 22-4)

⌒5.517

1.000

82°

2.875

5.000

| Your Name Here | Arc Length with Orthogonal Ext. | Ex-22L |

EXERCISE 22M

INSTRUCTIONS:
1. Open **Border A**
2. Draw objects below using Circles, Arcs, Lines and Fillet on layer Object Line.
3. Dimension as shown using Dimension Style **Class Style** and Layer Dimension.
   (Refer to page 18-3 and 18-5 for Radius and Diameter dimension settings)
4. Edit the Title and Ex-XX by double clicking on the text. Do not erase and replace.
5. Save as **EX22M**
6. Plot using Page Setup **Class Model A**

Refer to the next page for drawing assistance.

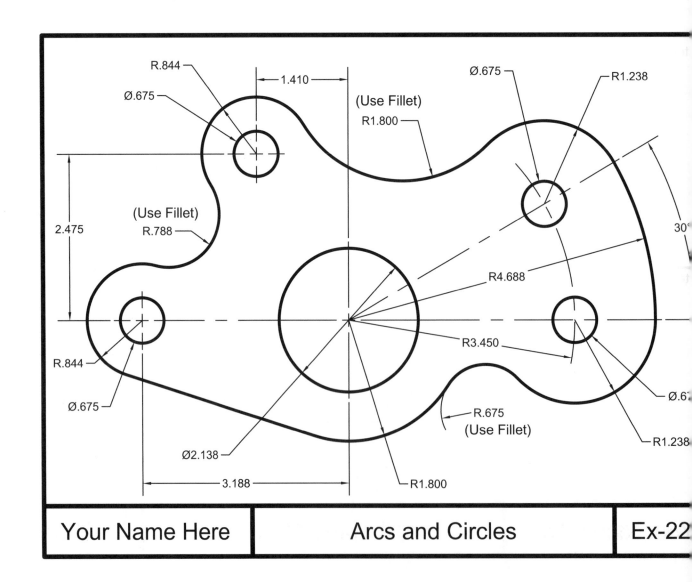

| Your Name Here | Arcs and Circles | Ex-22 |

EX 22M Helper

INSTRUCTIONS:
Step 1.
-Draw the horizontal and vertical lines then offset to create intersections for circles.
-Draw the 30 degree line

Step 2.
-Draw the Circles

Step 3.
-Draw the Arcs and draw the circles.

Step 4.
-Fillet, draw the tangent line and dimension.

EXERCISE 22N

INSTRUCTIONS:
1. Open **Border A**
2. Draw a 9" Radius Circle on Layer Object line, as shown below.
3. Trim so it does not extend beyond the the border lines.
4. Dimension as shown using Dimension Style **Class Style** and Layer Dimension. (Refer to page 22-5)
5. Edit the Title and Ex-XX by double clicking on the text. Do not erase and replace.
6. Save as **EX22N**
7. Plot using Page Setup **Class Model A**

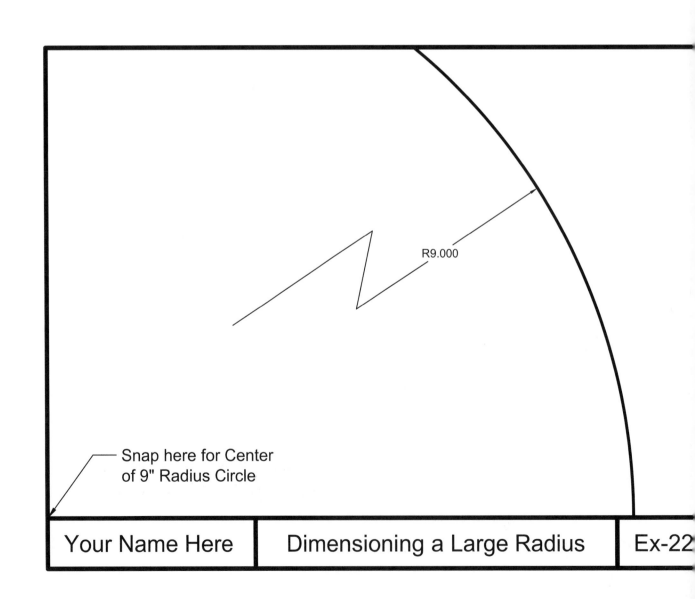

R9.000

Snap here for Center
of 9" Radius Circle

| Your Name Here | Dimensioning a Large Radius | Ex-22 |

LEARNING OBJECTIVES

After completing this lesson, you will be able to:

1. Understand the Polyline command
2. Draw a Polyline and a Polyarc
3. Assign widths to polylines
4. Set the Fill mode to On or Off.
5. Explode a Polyline

LESSON 23

POLYLINES

A **POLYLINE** is very similar to a LINE. It is created in the same way a line is drawn. It requires first and second endpoints. But a POLYLINE has additional features, as follows:

      1. A **POLYLINE** is ONE object, even though it may have many segments.
      2. You may specify a specific width to each segment.
      3. You may specify a different width to the start and end of a polyline segment.

Select the Polyline Command using one of the following:

      Ribbon = Home tab / Draw panel /

      Keyboard = PL <enter>

THE FOLLOWING ARE EXAMPLES OF POLYLINES WITH WIDTHS ASSIGNED.

Start and End widths .500

Start width 1.00 End width 0

2 Polyline segments

End width .25

Start width 1.00

Single Polyline Arc

Refer to the next page for more polyline options

POLYLINES....continued

OPTIONS:

WIDTH
Specify the starting and ending width.

You can create a tapered polyline by specifying different starting and ending widths.

HALFWIDTH
The same as Width except the starting and ending halfwidth specifies half the width rather than the entire width.

ARC
This option allows you to create a circular polyline less than 360 degrees. You may use (2) 180 degree arcs to form a full circular shape.

CLOSE
The close option is the same as in the Line command. Close attaches the last segment to the first segment.

LENGTH
This option allows you to draw a polyline at the same angle as the last polyline drawn. This option is very similar to the OFFSET command. You specify the first endpoint and the length. The new polyline will automatically be drawn at the same angle as the previous polyline.

POLYLINES....continued

CONTROLLING THE FILL MODE

If you turn the FILL mode OFF the polylines will appear as shown below.

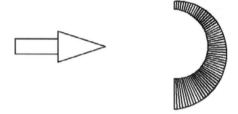

How to turn FILL MODE on or off.
1. Command: *type* **FILL** *<enter>*
2. Enter mode [On / Off] <ON>: *type ON or Off <enter>*
3. Command: *type REGEN <enter> or select: View / Regen*

EXPLODING A POLYLINE

NOTE: If you explode a POLYLINE it loses its width and turns into a regular line as shown below.

POLYLINES….continued

The following is an example of how to draw a polyline with Width.

1. Select the Polyline Command using one of the following:

Ribbon = Home tab / Draw panel /

Keyboard = PL <enter>

    Command: _pline

2. Specify start point: ***place the first endpoint of the line***

    Current line-width is 0.000
3. Specify next point or [Arc/Halfwidth/Length/Undo/Width]: ***select width option***

4. Specify starting width <0.000>: ***1*** <enter>

5. Specify ending width <0.000>: ***1*** <enter>

6. Specify next point or [Arc/Close/Halfwidth/Length/Undo/Width]: ***3 <enter>***

7. Specify starting width <0.000>: ***2*** <enter>

8. Specify ending width <0.000>: ***.5*** <enter>

9. Specify next point or [Arc/Close/Halfwidth/Length/Undo/Width]: ***2.75 <enter>***

10. Specify starting width <0.000>: ***.5*** <enter>

11. Specify ending width <0.000>: ***1*** <enter>

12. Specify next point or [Arc/Close/Halfwidth/Length/Undo/Width]: ***2.50 <enter>***

EXERCISE 23A

INSTRUCTIONS:
1. Open **Border A**
2. Draw the Polylines as shown using layer Object line.
3. Do not dimension
4. Edit the Title and Ex-XX by double clicking on the text. Do not erase and replace.
5. Save as **EX23A**
6. Plot using Page Setup **Class Model A**

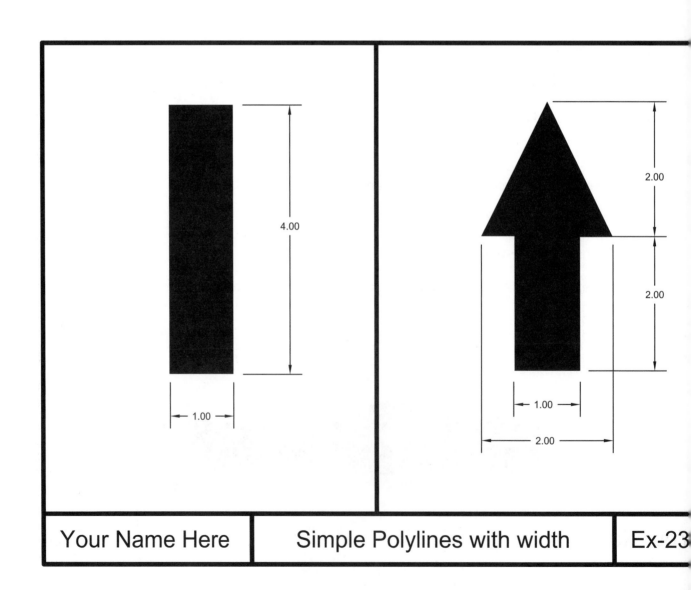

| Your Name Here | Simple Polylines with width | Ex-23 |

EXERCISE 23B

INSTRUCTIONS:

1. Open **EX23A**
2. Turn off the **Fill** mode. (Refer to page 23-4)
3. Do not dimension
4. Edit the Title and Ex-XX by double clicking on the text. Do not erase and replace.
5. Save as **EX23B**
6. Plot using Page Setup **Class Model A**

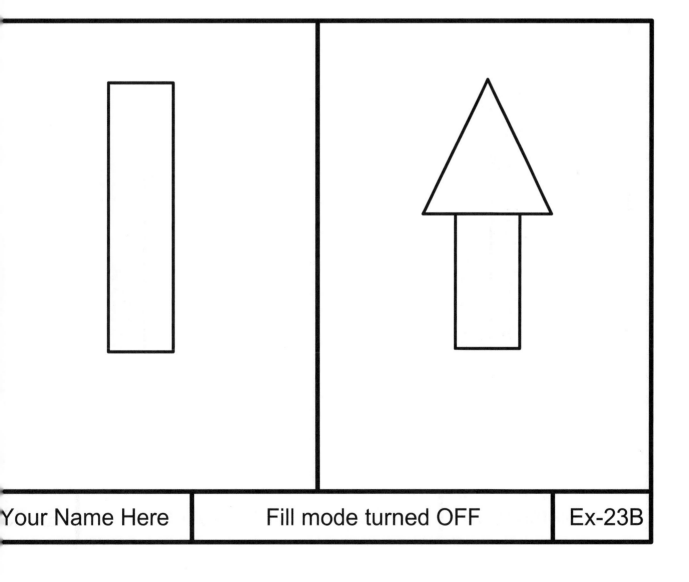

| Your Name Here | Fill mode turned OFF | Ex-23B |

EXERCISE 23C

INSTRUCTIONS:
1. Open **EX23B**
2. Explode each of the polylines.

   Note: I am sure most Architectural students were thinking that the polyline command would be good to use when creating Floor plan walls. But notice what happens if you explode the polylines. Consider Multiline or Double line (LT) commands for Floor plan walls. You can find these commands in the Help Menu

3. Edit the Title and Ex-XX by double clicking on the text. Do not erase and replace.
4. Save as **EX23C**
5. Plot using Page Setup **Class Model A**

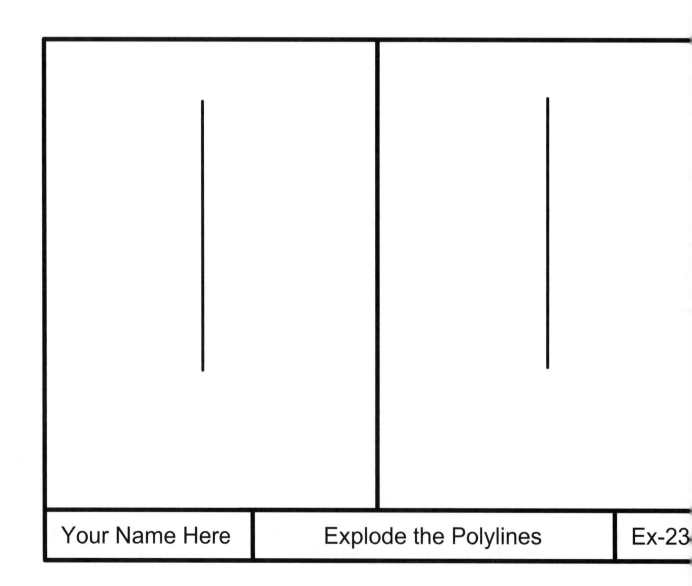

Your Name Here | Explode the Polylines | Ex-23

EXERCISE 23D

INSTRUCTIONS:
1. Open **Border A**
2. Draw the Polyline as shown using layer Object Line.
3. Do not Dimension.
4. Edit the Title and Ex-XX by double clicking on the text. Do not erase and replace.
5. Save as **EX23D**
6. Plot using Page Setup **Class Model A**

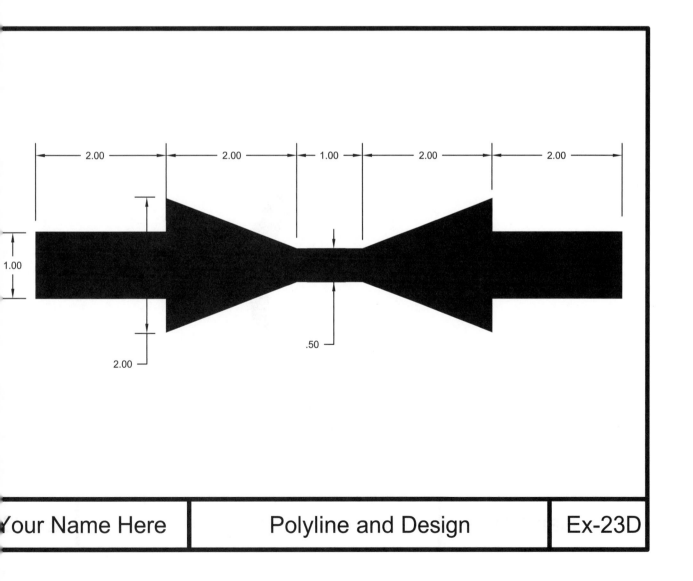

| Your Name Here | Polyline and Design | Ex-23D |

EXERCISE 23E

INSTRUCTIONS:
1. Open **Border A**
2. Select the Polyline Command.
3. Place the **start point** approximately as shown.
4. Set the **W**idths: Starting width = .10   Ending width = .50
5. Select option **A**rc
6. Select option **R**adius and enter **3.50**
7. Select option **A**ngle and enter **180**
8. Enter **90** for the **Direction of the Chord.**
9. Press **<enter>** to <u>stop</u>
10. Do not Dimension.
11. Edit the Title and Ex-XX by double clicking on the text. Do not erase and replace.
12. Save as **EX23E**
13. Plot using Page Setup **Class Model A**

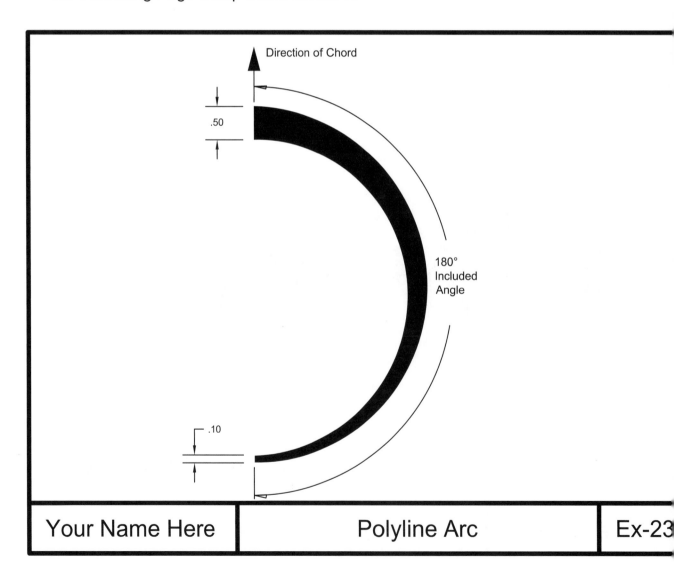

| Your Name Here | Polyline Arc | Ex-23 |

LEARNING OBJECTIVES

After completing this lesson, you will be able to:

1. Edit the width of Polylines
2. Join Polylines
3. Convert Polylines to Curves
4. Convert a basic Line to a Polyline
5. Join Lines, Polylines, Arcs and Splines

LESSON 24

EDITING POLYLINES

The **POLYEDIT** command allows you to make changes to a polyline's option, such as the width. You can also change a regular line into a polyline and JOIN the segments.

Note: If you select a line that is **NOT a POLYLINE**, the prompt will ask if you would like to turn it into a POLYLINE.

1. Select the **POLYEDIT** command using one of the following:

   Ribbon = Home tab / Modify ▼ panel /

   Keyboard = PE <enter>

   Note: You may modify "Multiple" polylines simultaneously.

2. PEDIT Select polyline or [Multiple]: *select the polyline to be edited or "M"*

3. Enter an option [Close/Join/Width/Edit vertex/Fit/Spline/Decurve/Ltypegen/Undo]: *select an Option (descriptions of each are listed below.)*

OPTIONS

CLOSE
CLOSE connects the last segment with the first segment of an Open polyline. AutoCAD considers a polyline open unless you use the "Close" option to connect the segments originally.

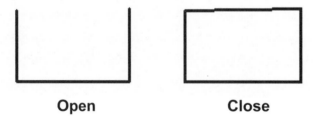

Open	**Close**

OPEN
OPEN removes the closing segment, but only if the CLOSE option was used to close the polyline originally.

EDITING POLYLINES....continued

OPTIONS:

JOIN
The JOIN option allows you to join individual polyline segments into one polyline.
The segments must have matching endpoints.

(4) Individual Polylines   **Joined
1 Polyline**

WIDTH
The WIDTH option allows you to change the width of the polyline.
But the entire polyline will have the same width.

EDIT VERTEX
This option allows you to change the starting and ending width of each segment individually.

SPLINE
This option allows you to change straight polylines to curves.

Polyline   **Splined**

DECURVE
This option removes the SPLINE curves and returns the polyline to its original straight line segments.

Splined   **Decurve**

JOIN COMMAND

I know this might seem confusing but this Join command is not the same as the Join "option" within the Polyedit tool. The Polyedit option can only be used to join individual polyline segments. This **Join** Command joins similar objects to form a single unbroken object.

1. Select the JOIN command using one of the following:

    Ribbon = Home tab / Modify ▾ panel / ┃ ⊣⊢ ┃

    Keyboard = J <enter>

2. Select the **Source** object.

3. Select the **objects to join to the source** object.

Seems easy, and it is, but there are a few rules regarding each type of object you must understand.

LINES
The Lines must be collinear (on the same plane) but can have gaps between them.

Before Join   **After Join**

POLYLINES
Same rules as Lines except no gaps allowed between them.

Before Join   **After Join**

ARC
The Arcs must lie on the same imaginary circle but can have gaps.

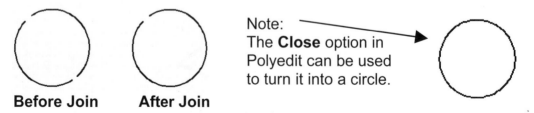

Before Join   **After Join**

Note:
The **Close** option in
Polyedit can be used
to turn it into a circle.

SPLINE
The Spline must lie in the same plane, and contiguous (end to end).
(You haven't learned this command as yet)

EXERCISE 24A

INSTRUCTIONS:
1. Open **Border A**
2. Draw the Polyline as shown using layer Object line.
3. Using the Polyedit command, Close and Open the polyline.
4. Edit the Title and Ex-XX by double clicking on the text. Do not erase and replace.
5. Do not dimension
6. Save as **EX24A**
7. Not necessary to plot.

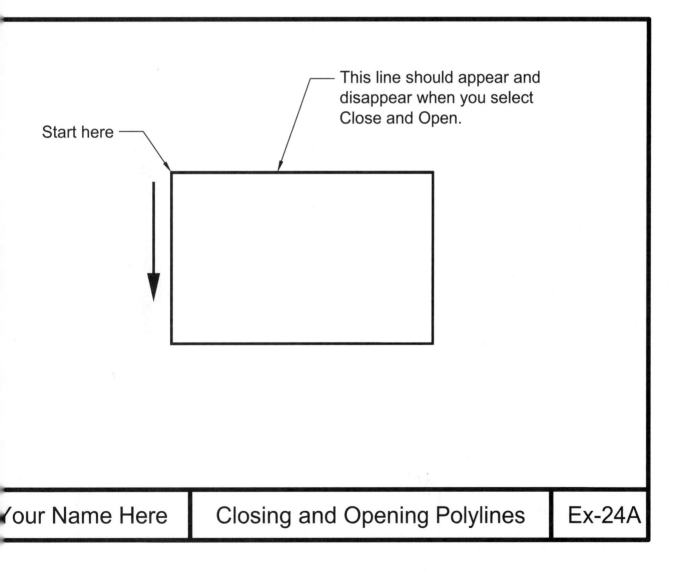

This line should appear and disappear when you select Close and Open.

Start here

Your Name Here | Closing and Opening Polylines | Ex-24A

EXERCISE 24B

INSTRUCTIONS:
1. Open **Border A**

Step 1.
2. Draw the Polyline 4" long and 0 width using layer Object line.

Step 2
3. Change the width of the Polyline to .50
4. Do not dimension
5. Edit the Title and Ex-XX by double clicking on the text. Do not erase and replace.
6. Save as **EX24B**
7. Plot using Page Setup **Class Model A**

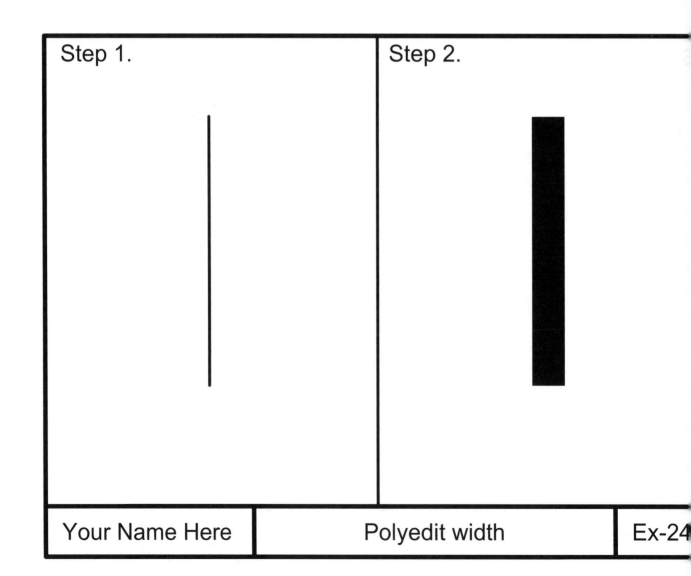

Step 1.	Step 2.	
Your Name Here	Polyedit width	Ex-24

EXERCISE 24C

INSTRUCTIONS:
1. Open **Border A**
Step 1.
2. Draw 1 continuous polyline with (3) 2" Long segments.
Step 2
3. Select the Polyedit command and select the first segment.
4. Select the Edit Vertex option. (Notice the "X" marking the starting point)
5. Select the Width option and set Start to 1.00 and Ending to .50.
6. Select Next option (The "X" should have moved to the next segment.)
7. Select the Width option and set Start to .50 and Ending to .50
8. Select the Next option (The "X" should have moved to the next segment)
9. Select the Width option and set Start to 2.00 and Ending to 0
10. Select the Exit option and press <enter> to stop
11. Edit the Title and Ex-XX ;   Save as **EX24C;**   Plot using Page Setup **Class Model A**

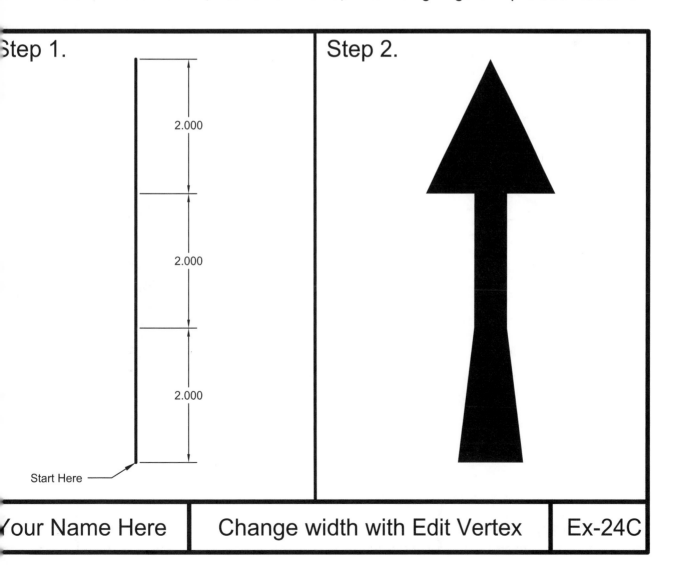

Step 1.	Step 2.	
Your Name Here	Change width with Edit Vertex	Ex-24C

EXERCISE 24D

INSTRUCTIONS:

1. Open **Border A**

Step 1.

2. Draw the objects below using the **LINE** command **not polyline**.

Step 2

3. Select the Polyedit command and select the first segment drawn.
   (Answer **YES** to "*Not a polyline, do you want to turn it into one*?")
4. Select the Width option and set the width to .10 (Notice only one segment changed)
5. Select the Join option.
6. Select the remaining lines using a crossing window.
   (Now the width should be constant for all lines)
7. Edit the Title and Ex-XX
8. Save as **EX24D**
9. Plot using Page Setup **Class Model A**

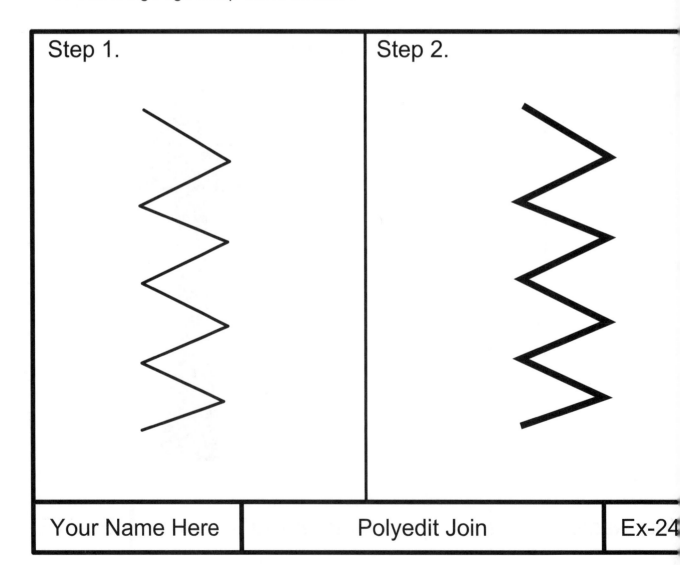

Step 1.   Step 2.

Your Name Here   Polyedit Join   Ex-24

EXERCISE 24E

INSTRUCTIONS:
1. Open **EX24D**
2. Select the Polyedit command
3. Select the Polyline
4. Select the Spline option
5. Edit the Title and Ex-XX
6. Save as **EX24E**
7. Plot using Page Setup **Class Model A**

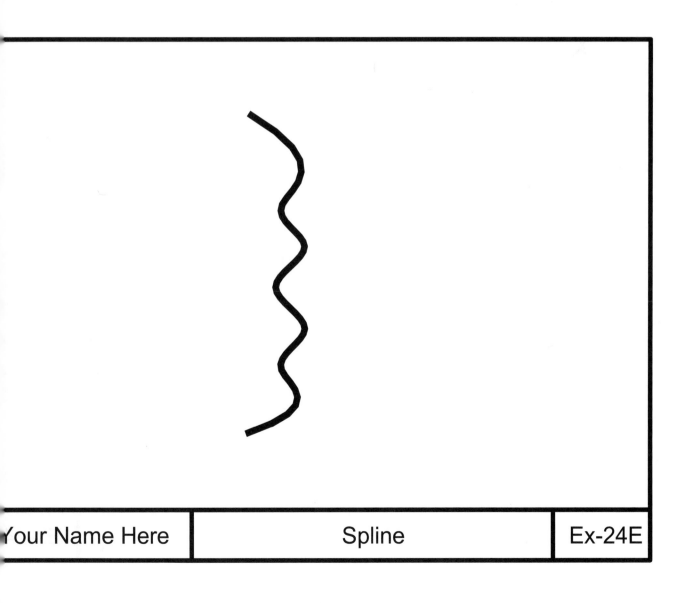

Your Name Here	Spline	Ex-24E

Notes:

LEARNING OBJECTIVES

After completing this lesson, you will be able to:

1. Create a new Text Style
2. Change an existing Text Style
3. Delete a Text Style
4. Select a Point Style
5. Divide an object into specified number of segments
6. Divide an object to a specified measurement

LESSON 25

CREATING NEW TEXT STYLES

AutoCAD provides you with two preset Text Styles named "Standard" and "Annotative". (Annotative will be discussed in Lesson 27)
You may want to create a new text style with a different font and effects.
The following illustrates how to create a new text style.

1. Select the TEXT STYLE command using one of the following:

   Ribbon = Annotate tab / Text panel / ↘

   Keyboard = ST <enter>

2. Select the **NEW** button.

3. Type the new style name in **STYLE NAME** box.
   Styles names can have a maximum of 31 characters, including letters, numbers, dashes, underlines and dollar signs. You can use Upper or Lower case.

4. Select the **OK** button.

5. Select the **FONT**.

6. Enter the value of the Height.
   Note: If the value is 0, AutoCAD will always prompt you for a height.
   If you enter a number the new text style will have a fixed height and AutoCAD will not prompt you for the height.

CREATING NEW TEXT STYLES....continued

7. Assign **EFFECTS.**

UPSIDE-DOWN
Each letter will be created upside-down in the order in which it was typed.
(Note: this is different from rotating text 180 degrees.)

BACKWARDS
The letters will be created backwards as typed.

VERTICAL
Each letter will be inserted directly under the other. Only **.shx** fonts can be used.
VERTICAL text will not display in the **PREVIEW** box.

OBLIQUE ANGLE
Creates letter with a slant, like italic. An angle of 0 creates a vertical letter. A
positive angle will slant the letter forward. A negative angle will slant the letter
backward.

WIDTH FACTOR
This effect compresses or extends the width of each character.
A value less than 1 compresses each character.
A value greater than 1 extends each character.

8. **PREVIEW** your settings

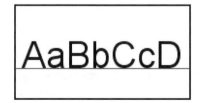

9. Select the **Set Current** button.

10. Select the **Apply** button

HOW TO SELECT A TEXT STYLE

After you have created Text Styles you will want to use them. You must select the Text Style before you use it.

Below are the methods when using Single Line or Multiline text.

Single Line Text

<u>Select the style before selecting Single Line Text command.</u>

1. Select the **Annotate** tab.
2. Using the Text panel, select the style down arrow ▾
3. Select the Text Style

Multiline Text

<u>Select the Text Style within the Text Editor.</u>

1. Find the **Style** panel and scroll through the text styles available using the up and down arrows.

DELETE A TEXT STYLE

1. Select the **TEXT STYLE** command using one of the following:

   Ribbon = Annotate tab / Text panel / ↘

   Keyboard = ST <enter>

2. First, select a **Text Style** that you <u>do not want to Delete</u> and select **Set Current** button. (You can't Delete a Text Style that is in use.)

3. Select the **Text Style** that you want to Delete.

4. Select the **Delete** button.

5. Warning appears, select **Ok** or **Cancel**

6. Select the **Close** button.

Note: Also refer to page 29-9 for the PURGE command. The Purge command will remove any unused text styles, dimension styles, layers and linetypes.

CHANGE EFFECTS OF A TEXT STYLE

1. Select the **TEXT STYLE** command using one of the following:

   Ribbon = Annotate tab / Text panel / ↘

   Keyboard = ST <enter>

2. Select the **Text Style** you wish to change.

3. Make the changes in the **EFFECTS** boxes.

   Note about Vertical:
   Only **.shx** fonts can be vertical
   Vertical text will not display in the **PREVIEW** box.

4. Select the **Apply** button. (*Apply will stay gray if you did not change a setting.*)

5. Select the **Close** button.

DIVIDE COMMAND

The DIVIDE command divides an object mathematically by the NUMBER of segments you specify. A POINT (object) is placed at each interval on the object.

Note: the object selected is **NOT** broken into segments. The **POINTS** are simply drawn **ON** the object.

EXAMPLE:

This LINE has been DIVIDED into 4 EQUAL lengths.
But remember, the line is not broken into segments.
The Points are simply drawn **ON** the object.

1. First open the Point Style box and select the **POINT STYLE** to be placed on the object.

   Keyboard = ddptype <enter>

   Select either of these ONLY when you want to make the points disappear yet still be there.

(Refer to page 5-7 if you need a refresher on Points)

2. Next select the **DIVIDE** command using one of the following:

   Ribbon = Home tab / Draw panel ▼/ [icon] (on the Point tool)

   Keyboard = DIV <enter>

3. Select Object to divide: *select the object to divide*

4. Enter the number of segments or [Block]: *type the number of segments <enter>*

MEASURE COMMAND

The **MEASURE** command is very similar to the **DIVIDE** command because point objects are drawn at intervals on an object. However, the **MEASURE** command allows you to specify the **LENGTH** of the segments rather than the number of segments.

Note: the object selected is **NOT** broken into segments. The **POINTS** are simply drawn **ON** the object.

EXAMPLE:

Select Object here to start the measurement From this end.

The **MEASURE**ment was started at the left endpoint, and ended just short of the right end of the line. The remainder is less than the measurement length specified. You designate which end you want the measurement to start by selecting the end when prompted to select the object.

1. First open the Point Style box and select the **POINT STYLE** to be placed on the object.

    Keyboard = ddptype <enter>

(Refer to page 5-7 if you need a refresher on Points)

2. Next select the **MEASURE** command using one of the following:

    Ribbon = Home tab / Draw panel ▼/ ⟨ (on the Point tool)

    Keyboard = ME <enter>

3. Select Object to MEASURE: *select the object to MEASURE*
    (Note: this selection point is also where the MEASUREment will start.)

4. Specify length of segment or [Block]: *type the length of one segment <enter>*

EXERCISE 25A

INSTRUCTIONS:

1. Open **Border A**

2. Create the 3 Text Styles shown below. (Style1, Style2 and Style3)

   Select all of the options shown.

3. **Save** the drawing as: **EX25A**

STYLE1

Note: Style2 and Style3 on the next page.

EXERCISE 25A.....continued

STYLE1

STYLE 2

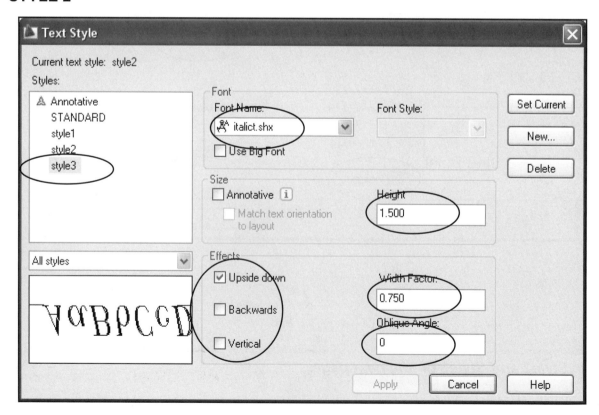

EXERCISE 25B

INSTRUCTIONS:

1. Open **EX25A**
2. Before entering text, move the Origin to the lower left corner as shown.
3. Draw the 3 Text Styles using "Single Line Text".

   Remember to select the text style before selecting the Single Line text command.

4. Place the start point exactly as shown. (enter coordinates)
5. Use layer Text.
6. Edit the Title and Ex-XX by double clicking on the text. Do not erase and replace.
7. Save as **EX25B**
8. Plot using Page Setup **Class Model A**

Start Point 5.20,5.90

Start Point 2.60,5.50

Start Point .90. 2.17

Move Origin here before entering text.

| Your Name Here | Using the new Text styles | Ex-25B |

EXERCISE 25C

INSTRUCTIONS:

1. Open **EX25B**
2. Change the Text to Text Style "Standard".
   a. Select all text
   b. Right click and select "Properties" from the pop up menu.
   c. Select Style "Standard"

   (Note: You may also use the **Quick Properties** box.)

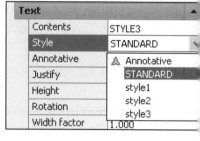

3. Edit the Title and Ex-XX by double clicking on the text.
4. Save as **EX25C**
5. Plot using Page Setup **Class Model A**

STYLE1 STYLE2

STYLE3

| Your Name Here | Changing the Text styles | Ex-25 |

EXERCISE 25D

INSTRUCTIONS:

1. Open **EX25C**

2. Delete Text Styles 1, 2 and 3 as follows:

    a. Select **TEXT STYLE box.**

    b. Select the "Standard" text style.

    c. Select **Set Current** button. (This makes Standard the current text style)

        AutoCAD will not allow you to delete an active or current text style. That is why you must make the "Standard" text style current before you may delete a text style.

    d. Select the style you want to **delete**.

    e. Click on the **DELETE** button.

3. **Save** as: **EX25D**

EXERCISE 25E

INSTRUCTIONS:

1. Open **Border A**
2. Draw the (2) 8" long lines shown below on layer Object line.
3. **Divide** the upper line into 10 equal segments.
4. Using **Measure** create (5) 1.50 long segments on the lower line as shown.
5. Edit the Title and Ex-XX by double clicking on the text. Do not erase and replace.
6. Dimension as shown. (Remember "Node" is the object snap to use for Points)
7. Save as **EX25E**
8. Plot using Page Setup **Class Model A**

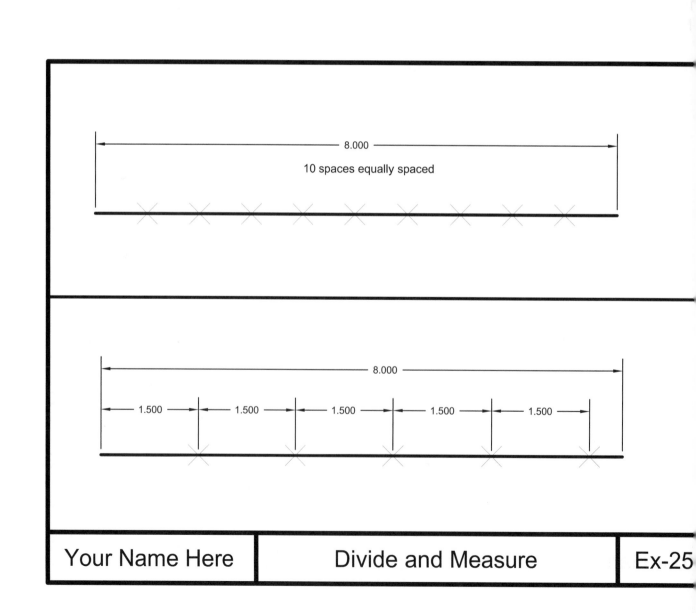

| Your Name Here | Divide and Measure | Ex-25 |

25-14

EXERCISE 25F

INSTRUCTIONS:

1. Open **Border A**
2. Draw the "pretend" house roof plan with fence line shown below.
3. Use Layers Roof, Walls and Hidden lines.
4. Divide the Fence lines as shown. (Important: If you used Rectangle, explode it)
5. Draw 1 Fence Post, on layer Object line, at upper left corner using Inscribed, 4 sided, .10 radius Polygon.
6. Copy the Fence Post to the "Points" and corners using object snap "Node" and "Endpoint".
7. Select the Point Style that does not display a point. (Refer to page 25-7)
8. Dimension using Dimension Style "Class Style" and Layer Dimension
9. Edit the Title and Ex-XX by double clicking on the text. Do not erase and replace.
10. Save as **EX25** and Plot using Page Setup **Class Model A**

(Note: You will learn to actually draw a full size house in Lesson 27.)

| Your Name Here | Pretend House with Fence posts | Ex-25F |

Notes:

LEARNING OBJECTIVES

After completing this lesson, you will be able to:

1. Understand the difference between Model and Layout tabs
2. Create Viewports
3. Create a Page Setup
4. Create a Plot Page Setup
5. Create a Decimal Setup Master template
6. Create a new Border
7. Experiment with multiple viewports

LESSON 26

SERIOUS BUSINESS

In the previous lessons you have been having fun learning most of the basic commands that AutoCAD offers and you have been using a template that included preset layers and drawing settings etc. But now it is time to get down to the "serious business" of <u>setting up your own drawing from "scratch"</u>.

<u>Starting from scratch</u> means you will need to set or create the following:

<u>Items 1 through 5 you have learned in previous lessons</u>
1. Drawing Units (Lesson 4)
2. Snap and Grid (Lesson 2)
3. Create Text styles (Lesson 25)
4. Create Dimension Styles (Lesson 16)
5. Create new layers and load linetypes.

<u>Items 6 through 9 will be learned in this lesson</u>
6. Create a "Layout" for plotting.
7. Create a "Floating Viewport" in the Layout.
8. Create a "Page Setup" to save plot settings.
9. Plot the drawing from Paper space.

After reading pages 26-3 through 26-22 start Exercise 26A and work your way through to 26D. When you have completed Exercise 26E you will have created a master template named, "**<u>My Decimal Setup</u>**". This master template will have everything set, created and prepared, ready to use each time you want to create a drawing using decimal units and to be plotted on an 8-1/2 X 11 inch sheet.

This means, for future drawings you merely select **File / New** and "**My Decimal Setup.dwt**" and start drawing. No time consuming setups. It is all ready to go.

In Lesson 27 you will create a master template for "feet and inches".

So take it one page at a time and really concentrate on understanding the process.

Note:
I am using sheet size 8-1/2 X 11 so you may print the exercises on your printer. After you understand the page setup concept you will be able to assign a larger sheet size to conform to any large format printer that you may use in the future.

MODEL and LAYOUT OPTIONS

Very important:
*Before I discuss Model and Layout I need you to make a change to the options. The default AutoCAD screen does not display the **Model and Layout tabs**. AutoCAD gives you the option to displayed them or not. I prefer to have the tabs displayed. I think it will make it easier for you to understand.*

This will just take a minute.

1. Type **options <enter>**

2. Select the **Display** tab.

3. Check and un-check boxes as shown

3. Check and un-check boxes as shown.

4. Select the **OK** button

5. The lower left corner of the drawing area should now display the 3 tabs Model, Layout1 and Layout2 and a few new tools should be displayed in the lower right corner above the command line.

5. New

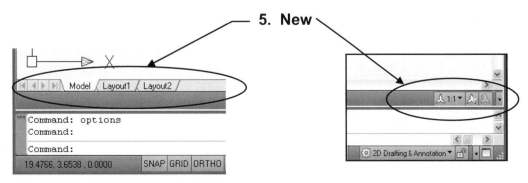

MODEL and LAYOUT tabs

Read this information carefully. It is very important that you understand this concept. More information on the following pages.

AutoCAD provides two drawing spaces, **MODEL** and **LAYOUT**. You move into one or the other by selecting either the MODEL or LAYOUT tabs, located at the bottom left of the drawing area. (If you do not have these displayed follow the instructions on the previous page.)

Refer to 26-3 if you do not have these.

Model Tab (Also called *Model Space*)

When you select the Model tab you enter <u>MODEL SPACE</u>.
(This is where you have been drawing and plotting from for the last 25 lessons)
Model Space is where you **create** and **modify** your drawings.

Layout Tabs (Also called *Paper Space*)

When you select a Layout tab you enter <u>PAPER SPACE</u>.
<u>The primary function of Paper Space is to</u> **prepare the drawing for plotting.**

When you select the Layout tab for the first time, the "<u>Page Setup Manager</u>" dialog box will appear. The Page Setup Manager allows you to specify the printing device and paper size to use.
(More information on this in "How to create a Page Setup" page 26-13)

When you select a Layout tab, Model Space will seem to have disappeared, and a <u>blank sheet of paper</u> is displayed on the screen. This sheet of paper is basically <u>in front of the Model Space</u>. (Refer to the illustration on the next page)

To see the drawing in Model Space, while still in Paper Space, you must <u>cut a hole</u> in this sheet. This hole is called a **"Viewport"**. *(Refer to "Viewports" page 26-6.)*

MODEL and LAYOUT tabs….continued

__Try to think of this as a picture frame (paper space) in front of a photograph (model space).__

Model Space
(Model tab)

Drawing in
model space

Paper Space
(Layout tab)

Border, title
block and notes
in paper space

Viewport Frame
(hole in paper space)

This is what you see when you select the ►‖ Model ⟋ tab.

You see only model space.

Viewport Frame

This is what you see when you select the ⟋ Layout1 ⟋ tab with a viewport.

You see through the Viewport to Model Space.

__Now hold this thought…….more explanation coming on the next few pages.__

VIEWPORTS

Note: This is just the concept to get you thinking. The actual step-by-step instructions will follow in the exercises.

Viewports are only used in Paper Space (Layout tab).
Viewports are holes cut into the sheet of paper displayed on the screen in Paper Space.
Viewports frames are objects. They can be moved, stretched, scaled, copied and erased. You can have multiple Viewports, and their size and shape can vary.

Note: It is considered good drawing management to create a layer for the Viewport "frames" to reside on. This will allow you to control them separately; such as setting the viewport layer to "No plot" so it will not be plot.

HOW TO CREATE A VIEWPORT

1. First, create a drawing in <u>Model Space (Model tab)</u> and save it.

2. Select the "Layout1" tab.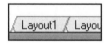

   When the "**Page Setup Manager**" dialog box appears, select the Modify button. Then you will select the <u>Printing device</u> and <u>paper size</u> to plot on. (Refer to "How to Create a Page setup" on page 26-13.)

3. You are now in <u>Paper Space</u>. Model Space appears to have disappeared, because a blank paper is now in front of Model Space, preventing you from seeing your drawing. You designated the size of this sheet in the "page setup" mentioned in #2 above. (The Border, title block and notes will be drawn on this paper.)

You designated the size of this sheet in the Page setup

The dashed line represents the printing limits for the printer that you selected in the Page setup.

Continued on the next page...

VIEWPORTS....continued

4. Draw a border, title block and notes in Paper Space (Layout)

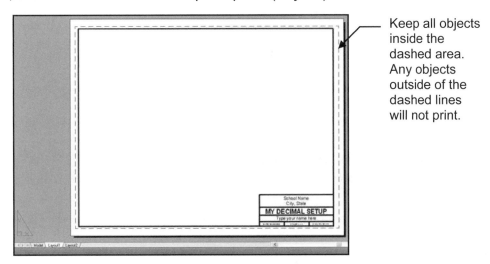

Keep all objects inside the dashed area. Any objects outside of the dashed lines will not print.

Now you will want to see the drawing that is in Model Space.

5. Select <u>layer</u> "**Viewport**"   (You want the viewport frame to be on layer viewport)

6. Type **MV <enter>**

7. Draw a Viewport "frame" by placing the location for the "first corner" and then the "opposite corner" using the cursor. (Similar to drawing a Rectangle, but **do not** use the Rectangle command. You must use the MV command)

Opposite Corner of Viewport

First corner of Viewport

You should now be able to look through the Paper Space sheet to Model Space and see your drawing because you just cut a rectangular shaped hole in the sheet.

Note: Now you may go back to Model Space or return to Paper Space, simply by selecting the tabs, model or layout.
(Make sure your grids are ON in Model Space and OFF in Paper Space. Otherwise you will have double grids)

WHY LAYOUTS ARE USEFUL

I know you are probably wondering why you should bother with Layouts.
A Layout (Paper Space) is a great method to manipulate your drawing for plotting.

Notice the drawing below with <u>multiple viewports</u>.

Each viewport is a hole in the paper.
You can see through each viewport (hole) to model space.

Using Zoom and Pan you can manipulate the display of model space in each viewport.
To manipulate the display you must be <u>inside</u> the viewport.
Refer to the next page for instructions on how to achieve this.

Note: the dashed line indicates the maximum printing area for the printer and paper selected. Any object outside of this area will not print.

Active viewport indicated by heavier viewport frame.

HOW TO REACH INTO A VIEWPORT

Here are the rules:
1. You have to be in Paper Space (layout tab) and at least one viewport must have been created.

2. You have to be <u>inside a Viewport </u>to manipulate the scale or position of the drawing that you see in that Viewport.

How to reach into a viewport to manipulate the display.

<u>First, select a layout tab and cut a viewport</u>

At the bottom middle of the screen on the status bar there is a button that either says Model or Paper. This button displays which space you are in currently.

When the button is PAPER you are working on the Paper sheet that is in front of Model Space. (Refer to the illustration on page 25 - 5)
You may cut a viewport, draw border, title block and place notes.

<u>If you want to reach into a viewport to manipulate the display</u>, click inside of the viewport frame. Only one viewport can be activated at one time. The active viewport is indicated by a heavier viewport frame. (Refer to the illustration on the previous page. The viewport displaying the doors is active).
Also, the Paper button changed to Model.

While you are inside a viewport you may manipulate the scale and position of the drawing displayed. To return to the Paper surface click on the word Model and it will change to Paper   You may now work on the paper surface.

*Note: Do not confuse the <u>**Model / Layout tabs** </u>with the <u>**MODEL / PAPER button**</u> .*

Here is the difference.

The **Model / Layout tabs** shuffle you from the actual drawing area (model space) to the Layout area (paper space). It is sort of like if you had 2 stacked pieces of paper and when you select the Model tab the drawing would come to the front and you could not see the layout. When you select the Layout tab a blank sheet would come to the front and you would not see model space......unless you have a viewport cut.

The **MODEL / PAPER button** allows you to work in model space or paper space without leaving the layout tab. No flipping of sheets. You are either on the paper surface or in the viewport reaching through to model space.

Don't worry it will get easier. This is the concept....but it will get more clear when you have completed the exercises in this lesson.

PAN

After you zoom in and out or adjust the scale of a viewport the drawing within the viewport frame may not be placed as you would like it. This is where **PAN** comes in handy. **PAN** will allow you to move the drawing around, within the viewport, without affecting the size.

Note: Do not use the MOVE command. You do not want to actually move the original drawing. You only want to slide the viewport image, of the original drawing, around within the viewport.

How to use the PAN command.

1. Select a layout tab (paper space)

2. Unlock the viewport if it is locked. (Refer to page 25-12)

3. Click inside a viewport.

4. Select the **PAN** command using one of the following:

    Ribbon = View tab / Navigate panel /

    Keyboard = P <enter>

    Status Line =

4. Place the cursor inside the viewport and hold the left mouse button down while moving the cursor. (Click and drag) When the drawing is in the desired location release the mouse button.

5. Lock the viewport.
   (Refer to page 26-12)

Refer to the Example on the next page.

PAN….continued

Before PAN

Click once in viewport to activate it.

To use PAN (Click, Drag, Release)

After PAN

HOW TO LOCK A VIEWPORT

After you have manipulated the drawing within each viewport, to suit your display needs, you will want to **LOCK** the viewport so the display can't be changed accidentally. Then you may zoom in and out and you will not disturb the display.

Note:
Accurately adjusting the scale of a viewport will be discussed in detail in Lesson 27 .

1. Make sure you are in **Paper Space**.

2. Click on a **Viewport Frame.**

Viewport Frame

3. Click on the **<u>Open Lock tool</u>** located in the lower right corner of drawing area.

The icon will change to a **<u>Closed Lock tool</u>**.

Viewport Unlocked   **Viewport Locked**

Now, any time you want to know if a Viewport is locked or unlocked just glance down to the <u>Lock tool</u> shown above.

Note: Do not use the Lock shown in the lower right corner of the screen.

This locks the size and position of toolbars and windows.

HOW TO CREATE A PAGE SETUP

When you select a layout tab for the first time the Page Setup Manager will appear. The Page Setup Manager allows you to select the **printer/plotter** and **paper size**. These specifications are called the "**Page Setup**". This page setup will be saved to that layout tab so it will be available when ever you use that layout tab.

Note: The following is for concept only. The actual exercise starts with 26A.

1. **Open** the drawing you wish to plot.
   (The drawing must be displayed on the screen.)

2. Select a **Layout tab**. (Refer to page 26-3 if you do not have a layout tab)

   Note: If the "Page Setup Manager" dialog box shown below does not appear automatically, right click on the Layout tab and select Page Setup Manager.

3. Select the **New...** button.

Check this box if the Page Setup Manager did not appear automatically when you selected a Layout tab.

Yours will be different. That's OK

4. Select **<Default output device>** in the Start with: list.

5. Enter the New page setup name: **Setup A**

6. Select **OK** button.

(I am assuming that your computer is attached to a printer. If not select Layout1)

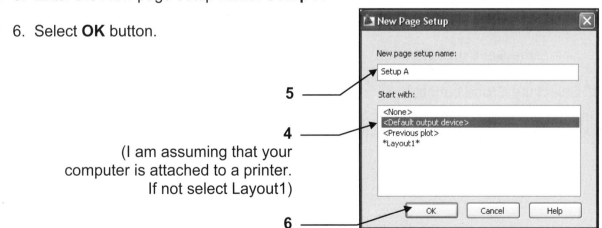

Continued on the next page...

HOW TO CREATE A PAGE SETUP....continued

This is where you will select the **printer / plotter**, **paper size** and the **plot offset**.

7. Select the **Printer / Plotter**
   Note: Your current system printer should already be displayed here. If you prefer another select the down arrow and select from the list. If the preferred printer is not in the list you must configure the printer. Refer to Appendix-A for instructions.)

8. Select the **Paper Size**

Notice the name you entered is now displayed as the page setup name.

9. Select **Plot Offset**

7
(Yours will be different)

8
(Yours may be 8-1/2 x 11)

9
(Should stay at 0.00000)

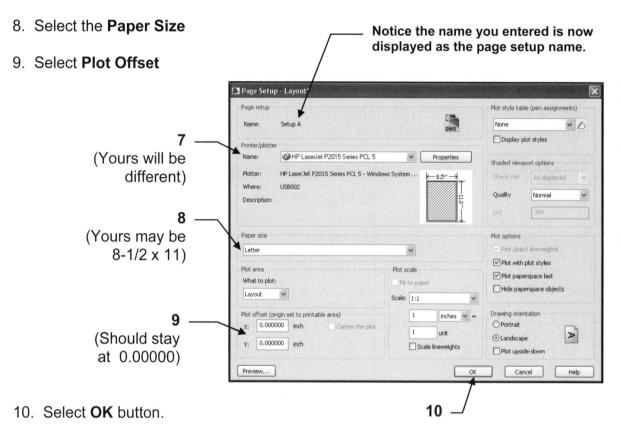

10. Select **OK** button.

10

11. Select the Page Setup.

12. Select the **Set Current** button.

13. Select the **Close** button.

11

12

13

HOW TO CREATE A PAGE SETUP....continued

You should now have a sheet of paper displayed on the screen.
This sheet is the size you specified in the "Page Setup".
This sheet is in front of Model Space.

The dashed line represents the maximum
printing area for the printing device that
you selected.
Any object outside of this area will not
print.

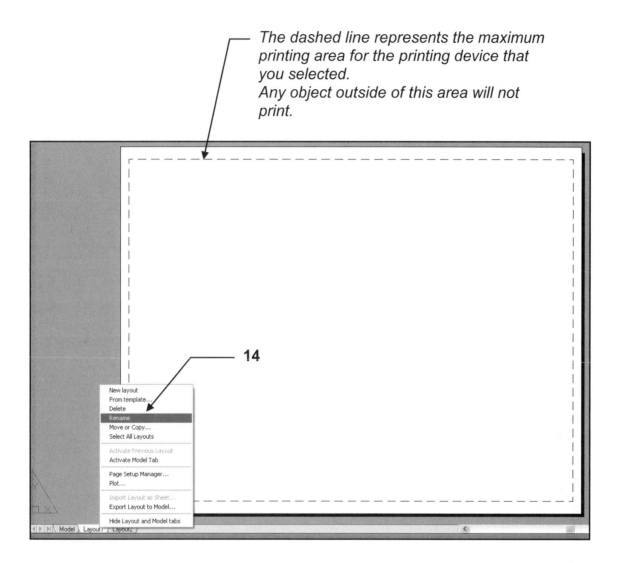

Rename the Layout tab

14. Right click on the active Layout tab and select **Rename** from the list.

15. Enter the new Layout name **A Size** <enter>

This was just the concept of Page Setup. You will actually create one in Exercise 26B.

USING THE LAYOUT

Now that you have the correct paper size on the screen, you need to do a little bit more to make it useful.

The next step is to:

1. Add a Border, Title Block and notes in Paper Space.

2. Cut a viewport to see through to Model Space.

3. Adjust the scale of the viewport as follows.
   A. You must be in Paper Space.
   B. Click on the Viewport Frame.
   C. Unlock Viewport if it is locked.
   D. Select the Viewport Scale down arrow ▼.
   E. Select Scale (List shown on next page)

Continued on the next page...

USING THE LAYOUT....continued

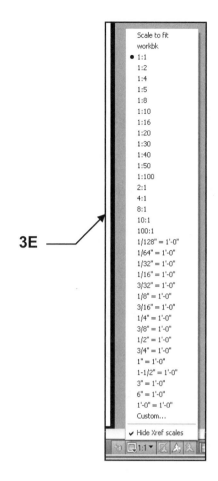

3E

4. Lock the Viewport.

Now you may zoom as much as you desire and it will not affect the adjusted scale.

Note: Adjusting the scale of the viewport will be discussed more in lesson 27.

HOW TO PLOT FROM THE LAYOUT

The previous page setup instructions were to select the printer and paper size.
Now you need to specify how you want to plot the drawing. You will find the PLOT
dialog box almost identical to the Page Setup dialog box.

1. Open the drawing you wish to plot.

2. Select the layout tab you wish to plot.

3. Select the Plot command using one of the following:

    Quick Access tool bar = 🖨

    Ribbon = Output tab / Plot panel / 🖨

    Application Menu = Print / Plot

    Keyboard = PLOT <enter>

The Plot dialog box shown below should appear.
Select the "More Options" ⊘ *button in the lower right corner*
if your dialog box does not appear the same as shown below

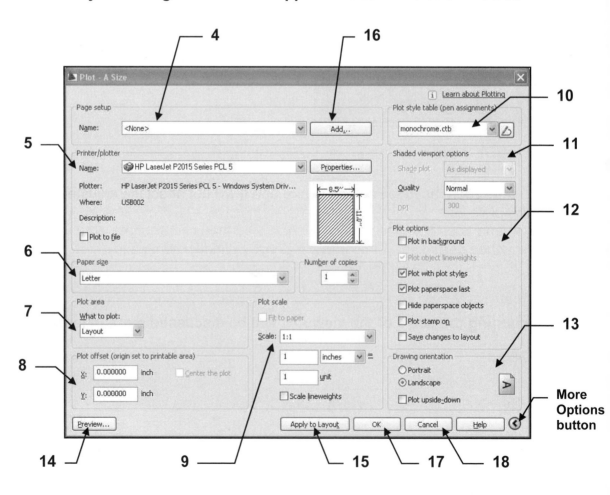

HOW TO PLOT FROM THE LAYOUT….continued

4. **Page Setup name**:
   After you have selected the desired settings you will save the new page setup and it will appear here. If you have previously created a page setup you may select it from the drop down list and all of the setting will change to reflect the previously saved page setup settings.

5. **Printer / Plotter:**
   Select the Printer that you wish to use. All previously configured devices will be listed here. (If your printer / plotter is not listed, refer to "Add a Printer / Plotter" Appendix A.)

6. **Paper Size:**
   Select the paper size. The paper sizes shown in the drop down list are the available sizes for the printer that you selected. If the size you require is not listed the printer you selected may not be able to handle that size. For example, a letter size printer can not handle a 24 X 18 size sheet. You must select a large format printer.

7. **Plot Area:**
   Select the area to plot. Layout is the default.

   Limits   plots the area inside the drawing limits.
   (This option is only available when plotting from model space)

   Layout   plots the paper size
   (Select this option when plotting from a Layout)

   Extents plots all objects in the drawing file even if out of view.
   (This option only available if you have a viewport)

   Display plots the drawing exactly as displayed on the screen.

   Window   plots objects inside a window. To specify the window, choose **Window** and specify the first and opposite (diagonal) corner of the area you choose to plot. (Similar to the Zoom / Window command)

8. **Plot offset:**
   The plot can be moved away from the lower left plot limit corner by changing the X and / or Y offset.
   If you have select **Plot area** "Display" or "Extents", select "**Center the plot.**"

9. **Scale:** Select a **scale** from the drop down list or enter a custom scale.

   *Note: This scale is the Paper Space scale. The Model space scale will be adjusted within the viewport. If you are plotting from a "**LAYOUT**" tab, normally you will use* **plot scale 1:1**. *(I know this seems a little confusing right now. Scaling will be discussed more in Lesson 27)*

HOW TO PLOT FROM THE LAYOUT….continued

10. **Plot Style Table:** Select the Plot Style Table from the list. The Plot Styles determine if the plot is in color, Black ink or screened. You may also create your own.
    If you want to print in Black Ink only select Monochrome.ctb
    If you want to print in Color select Acad.ctb

11. **Shaded viewport options**
    This area is used for printing shaded objects when working in the 3D environment.

12. **Plot options**
    Plot background = specifies that the plot is processed in the background.

    Plot Object Lineweights = plots objects with assigned lineweights.

    Plot with Plot Styles = plots using the selected Plot Style Table.

    Plot paperspace last = plots model space objects before plotting paperspace objects. Not available when plotting from model space.

    Hide Paperspace Objects = used for 3D only. Plots with hidden lines removed.

    Plot Stamp = Allows you to print information around the perimeter of the border such as; drawing name, layout name, date/time, login name, device name, paper size and plot scale.

    Save Changes to Layout = Select this box if you want to save all of these settings to the current Layout tab.

13. **Drawing Orientation.**
    Portrait = the short edge of the paper represents the top of the page.

    Landscape = the long edge of the paper represents the top of the page

    Plot Upside-down = Plots the drawing upside down.

14. Select **Preview** button.
    Preview displays the drawing as it will plot on the sheet of paper.

    (Note: If you cannot see through to Model space, you have not cut your viewport yet)

    If the drawing appears as you would like it, press the **Esc** key and continue.

    If the drawing does not look correct, press the **Esc** key and re-check your settings, then preview again.

Note: If you have any of the layers set to "no plot" they will not appear in the preview display. The Preview Display only displays what will be printed.

HOW TO PLOT FROM THE LAYOUT....continued

15. **Apply to Layout** button
This applies all of the settings to the layout tab. Whenever you select this layout tab the settings will already be set.

16. **Save the Page Setup**
At this point you have the option of saving these settings as another page setup for future use on other layout tabs. If you wish to save this setup, select the **ADD** button, type a name and select **OK.**

17. If your computer **is** connected to the plotter / printer selected, select the **OK** button to plot, then proceed to **19**.

18. If your computer is **not** connected to the plotter / printer selected, select the **Cancel** button to close the Plot dialog box and proceed to **19**.

   *Note: Selecting Cancel will cancel your selected setting if you did not save the page setup as specified in **16** above.*

19. **Save the drawing**
This will guarantee that the Page Setup you just created will be saved to this file for future use.

Note: This is the concept only.
The step by step instructions are shown in Exercise 26E

ANNOTATIVE PROPERTY

In Lesson 16 you learned to create a dimension style.
In Lesson 25 you learned to create a text style.
In this Lesson you will create a new dimension style and text style but this time you will include the **Annotative property** in both.

The Annotative property automates the process of scaling text, dimensions, hatch, tolerances, leaders and symbols. The height of these objects will automatically be determined by the annotation scale setting.

This will be discussed more in Lesson 27.
For now I just want you to know how to select it when creating your new styles.

TEXT STYLE

This symbol Indicates Annotative

Select the Annotative option here

DIMENSION STYLE

This symbol indicates Annotative

Select the Annotative option here

EXERCISE 26A

CREATE A MASTER DECIMAL SETUP TEMPLATE

The following instructions will guide you through creating a "Master" decimal setup template. The "2010-Workbook Helper" is an example of a Master setup template. Even though the screen appears blank, the actual file is full of settings, such as: Units, Drawing Limits, Snap and Grid settings, Layers, Text styles and Dimension Styles. Once you have created this "Master" template, you just open it and draw. No more repetitive inputting of settings.

NEW SETTINGS

A. Begin your drawing with a different template as follows:
   1. Select the **NEW** command..
   2. Select template file **acad.dwt** from the list of templates.
   3. Select **Open**

B. Set the drawing **Units** as follows:
   1. Type **UNITS <enter>**
   2. Change the **Type** and **Precision** as shown below.
   3. Select **OK** button

EXERCISE 26A continued

C. Set the **Drawing Limits** (Size of the drawing area) as follows:
  1. Type **Limits <enter>**
  2. Enter Lower left corner = 0.000, 0.000
  3. Enter Upper right corner = 11 , 8.5
  4. Use **" ZOOM / ALL"** to view the new limits
  5. Set your **Grids** to **ON** to display the limits (drawing area).

D. Set the Grids and Snap as follows:
  1. Type **DS<enter>**
  2. Select the **Snap and Grid** tab
  3. Change the settings as shown below.
  4. Select the **OK** button.

D2

Important:
Uncheck
Adaptive
Grid

D4

EXERCISE 26A continued

E. Change **Lineweight** settings as follows:
1. Right click on Lineweight button located on the status line LWT or +
2. Select **Settings**
3. Change to inches and adjust the Display scale as shown below.
4. Select the **OK** button.

F. Create **new Layers** as follows:
1. Load the linetypes listed below. (Refer to page 3-16 for instructions)
   Center2, Hidden, Phantom2
2. Assign names, colors, linetypes, lineweights and plotability as shown below:
   (Refer to 3-15 for instructions)

Notice these

S..	Name	O..	Freeze	Lock	Color	Linetype	Lineweight	Plot S...	Plot
	0				white	Continuous	Default	Color_7	
	BORDERLINE				white	Continuous	0.039"	Color_7	
	CENTERLINE				cyan	CENTER2	Default	Color_4	
	DIMENSION				blue	Continuous	Default	Color_5	
	GRADIENT				white	Continuous	0.039"	Color_7	
	HATCH				green	Continuous	Default	Color_3	
	HIDDEN LINE				magenta	HIDDEN	Default	Color_6	
✓	OBJECT LINE				red	Continuous	0.031"	Color_1	
	PHANTOM LINE				white	PHANTOM2	Default	Color_7	
	SECTION LINE				white	PHANTOM2	0.031"	Color_7	
	SYMBOL				white	Continuous	0.016"	Color_7	
	TEXT				blue	Continuous	Default	Color_5	
	THREADS				green	Continuous	Default	Color_3	
	VIEWPORT				green	Continuous	Default	Color_3	
	WINDOW				green	Continuous	Default	Color_3	
	WIPE OUT				white	Continuous	0.039"	Color_7	

G. Create a new **Text Style** as follows:
  1. Select **Text Style** (Refer to page 25-2)
  2. Create the text style **Text-Classic** using the settings shown below.
  3. When complete, select **Set Current** and **Close.**

H. Create a **Dimension Style** as follows:
  1. Follow the directions on page 16-9
  2. All settings will be the same except the following:
    <u>Name</u> = Dim-Decimal
    <u>Annotative</u> = this new dimension style will be Annotative (Refer to page 26-22)

EXERCISE 26A continued

I. Create a Dimension Sub-Style as follows:
1. Follow the directions on page 18-9
2. All settings will be the same except the following:
Annotative = this new sub-style will be Annotative (Refer to page 26-22)

Your Dimension Style Manager should now appear as shown below:

Make sure you have Annotative symbols

THIS NEXT STEP IS VERY IMPORTANT

J. **Save** all the settings you just created as follows:
1. Save as: **My Decimal Setup**

K. Now continue on to **Exercise 26B**....you are not done yet.

EXERCISE 26B

PAGE SETUP

Now you will select the printer and the paper size to use for printing.
You will use the Layout1 tab (Paper space).

A. Open **My Decimal Setup** (if not already open)

B. Select **Layout1** tab.

Refer to page 26-3 if you do not have these tabs.

Note: If the "Page Setup Manager" dialog box shown below does not appear automatically, right click on the Layout tab and select __Page Setup Manager__.

C. Select the **New...** button.

C

D. Select the **<Default output device>** in the Start with: list.

E. Enter the New page setup name: **Setup A**

F. Select **OK** button.

E

D

(I am assuming that your computer is attached to a printer. If not select Layout1)

F

Continued on the next page...

EXERCISE 26B continued

This is where you will select the **print device**, **paper size** and the **plot offset**.

G. Select the **Printer / Plotter**
   Note: Your current system printer should already be displayed here. If you prefer another select the down arrow and select from the list. If the preferred printer is not in the list you must configure the printer. Refer to Appendix-A for instructions.)

H. Select the **Paper Size**

I. Select **Plot Offset**

Notice the name you entered is now displayed as the page setup name.

G
(Yours will be different)

H
(Yours may be 8-1/2 x 11)

I
(Should stay at 0.00000)

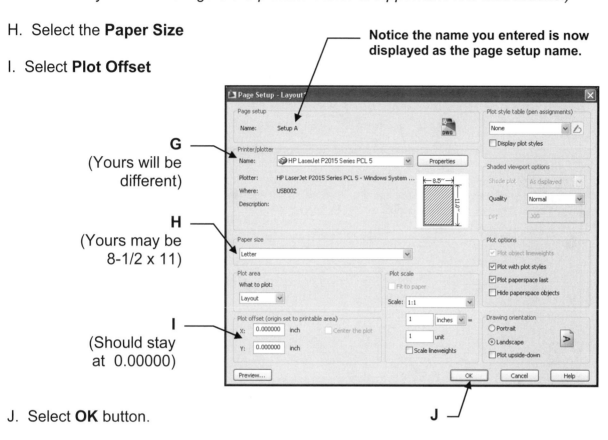

J. Select **OK** button.

J

K. Select **Setup A**.

L. Select the **Set Current** button.

M. Select the **Close** button.

K

L

M

Continued on the next page...

EXERCISE 26B continued

*You should now have a sheet of paper displayed on the screen.
This sheet is the size you specified in the "Page Setup".
This sheet is in front of Model Space.*

*The dashed line represents the maximum printing area for the printing device that you selected.
Any object outside of this area will not print.*

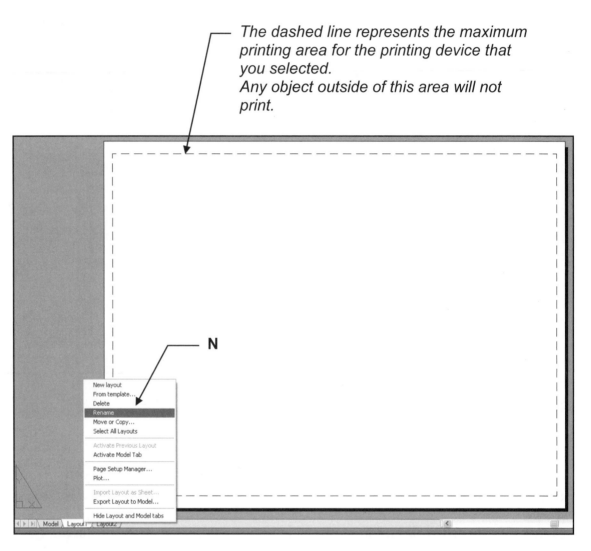

N

New layout
From template...
Delete
Rename
Move or Copy...
Select All Layouts

Activate Previous Layout
Activate Model Tab

Page Setup Manager...
Plot...

Import Layout as Sheet...
Export Layout to Model...

Hide Layout and Model tabs

Model ⟍ Layout1 ⟋ Layout2 ⟋

Rename the Layout tab

N. Right click on the **Layout1** tab and select **Rename** from the list.

O. Enter the New Layout name **A Size <enter>**

P. ***Very important:*** Save as **My Decimal Setup** <u>again</u>.

Q. Continue on to **Exercise 26C**.....you are not done yet.

EXERCISE 26C

CREATE A BORDER AND TITLE BLOCK

A. Draw the <u>Border rectangle</u> as large as you can within the dashed lines approximately as shown below using <u>Layer Border</u>.

B. Draw the Title Block as shown using:
1. Layers <u>Border</u> and <u>Text</u>.
2. Multiline Text ; Justify <u>Middle Center</u> in each rectangular area.
3. Text Style= Text-Classic   Text height varies.

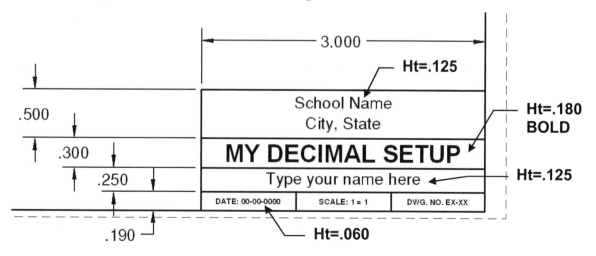

C. ***Very important:*** Save as **My Decimal Setup** again.

D. Continue on to **Exercise 26D**.....you are not done yet.

EXERCISE 26D

CREATE A VIEWPORT

The following instructions will guide you through creating a VIEWPORT in the Border Layout sheet. Creating a viewport has the same effect as cutting a hole in the sheet of paper. You will be able to see through the viewport frame (hole) to Modelspace.

A. Open **My Decimal Setup**
B. Select the **A Size** tab.
C. Select layer **Viewport**
D. Type: **MV <enter>**
E. Draw a Single viewport approximately as shown. (Turn off OSNAP)

Before   **After**

F. Adjust the Viewport Scale as follows:
   F1. Confirm "**Model" button** showing. MODEL
   F2. Select "**Zoom / All**"
   F3. Select the **Viewport Scale** of **1 = 1** (pg. 26-17)

G. Lock the Viewport (Refer to page 26-12)

H. *Very important....* Save as **My Decimal Setup** again.

I. Continue on to **Exercise 26E**....you are not done yet.

EXERCISE 26E

PLOTTING FROM THE LAYOUT

The following instructions will guide you through the final steps for setting up the master template for plotting. These settings will stay with **My Decimal Setup** and you will be able to use it over and over again.

A. Open **My Decimal Setup** (If not already open)

B. Select the **A Size** layout tab.

   You should be seeing your Border and Title Block now.

C. Select the **Plot** command

D. Select your printer

E. Select the Paper Size

F. Select the Plot Area

G. Plot offset should be 0.00000 for X and Y

EXERCISE 26E continued

H. Select scale 1 : 1

I. Select the Plot Style Table **Monochrome.ctb**

J. Select the Plot options shown

K. Select Drawing Orientation: **Landscape**

L. Select **Preview** button.

> If the drawing appears as you would like it, press the **Esc** key and continue
> on to **M**.

> If the drawing does not look correct, press the **Esc** key and re-check all the
> settings, then preview again.

Note: The Viewport will not appear in the Preview because it is on a no plot layer.

M. Select the **ADD** button.

N. Type the New page setup name: **Plot Setup A**

O. Select the **OK** button.

P. Select the "Apply to Layout" button.
    The settings are now saved to the layout tab for future use.

Q. If your computer **is** connected to the plotter / printer selected, select the **OK** button
    to plot, then proceed to **S**.

R. If your computer is **not** connected to the plotter / printer selected, select the Cancel
    button to close the Plot dialog box and proceed to **S**. *Note: Selecting Cancel will
    cancel your selected setting if you did not **ADD** the page setup as specified in **M**.*

S. You are almost done. Now Save all of this work as a **Template. (Refer to 2-3)**
    1. Select **Application Menu / Save As / AutoCAD Drawing Template**
    2. Type: **My Decimal Setup**
    3. Select **Save** button.

Wow, I know that seemed like a lot of work but you have now completed the
***Plotting Page Setup** for the **My Decimal Setup.dwt**.*
*Now you are ready to use this master template to create and plot many drawings in the
future. In fact, you have one on the very next page.*

EXERCISE 26F

This exercise is in 2 parts. First you will draw the drawing in the model tab. Then you will create a layout to display parts of the drawing in 3 viewports. *Hopefully this will help you understand the differences between model space and paper space.*

Part 1

A. Select **New** and select **My Decimal Setup** <u>template.</u>

B. Select the Model tab.

C. Draw the drawing shown below. **Do not dimension**.

D. Save as: **EX-26F1**.
    (Make sure that you save it as a drawing (.dwg) and not as a template.

Continue to Part 2 on the next page.

EXERCISE 26F continued

Part 2

A. Select the **A Size** tab.

B. Confirm the Model / Paper button says Paper.

C. Erase the one existing viewport frame. (Click on the frame and select Erase)

D. Select the Viewport Layer and create 3 new viewports approximately as shown below

E. Using Zoom and Pan, inside each viewport, try to display each viewport content as shown below.   (Note: Scale is not critical at this time but it will be in Lesson 27.)

F. Lock each viewport.

G. This time I would like to <u>print the Viewport Frames</u>, so you need to make the Viewpor layer **plottable**.
(Remove the slash 🖶 on the printer symbol in the Layer Properties Manager)

H. Change the Title block: Title: VIEWPORTS   Scale: NONE   Dwg no: EX-26F

I. Save as: **EX-26F**

J. Plot using Plot Page Setup name: **Plot Setup A**. (Remember to view the Preview)

LEARNING OBJECTIVES

After completing this lesson, you will be able to:

1. Understand Scaled Drawings
2. Adjust the scale within a Viewport
3. Understand Annotative Objects
4. Understand Paper space vs. Model Space dimensioning

LESSON 27

CREATING SCALED DRAWINGS

In the lessons previous to Lesson 26 you worked only in Model space. Then in Lesson 26 you learned that AutoCAD actually has another environment called Paper Space, or Layout. In this lesson we need to learn more about why we need 2 environments and how they make it easier to display and plot your drawings.

A very important rule in CAD you must understand is:

> **"*All objects are drawn full size*".**

In other words, if you want to draw a line 20 feet long, you actually draw it 20 feet long. If the line is 1/8" long, you actually draw it 1/8" long.

Drawing and Plotting objects that are very large or very small.

In the previous lessons you created medium sized drawings. Not too big, not too small. But what if you wanted to draw a house? Could you print it to scale on an 8-1/2 X 11 piece of paper? How about a small paper clip. Could you make it big enough to dimension? Let's start with the house.

How to print an entire house on an 8-1/2 X 11 sheet of paper.

Remember the photo and picture frame example I suggested in lesson 26? (Refer to 26-5) This time try to picture yourself standing at the front door of your house with an empty picture frame in your hands. Look at your house through the picture frame. Of course the house is way too big to fit in the frame. Or is it because you are standing too close to the house?

Now walk across the street and look through the picture frame in your hands again. Does the house appear smaller? Can you see all of it in the frame? If you could walk far enough away from the house it would eventually appear small enough to fit in the picture frame in your hands. But....the house did not actually change size, did it? It only appears smaller because you and the picture frame are farther away from it.

Adjusting the Viewport scale.

When using AutoCAD, walking across the street with the frame in your hands is called **Adjusting the Viewport scale.** You are increasing the distance between model space (your drawing) and Paper space (Layout) and that makes the drawing appear smaller. For example: A viewport scale of 1/4" = 1' would make model space appear 48 times smaller. But, when you dimension the house, the dimension values will be the actual measurement of the house. In other words, a 30 ft. line will have a dimension of 30'-0".

When plotting something smaller, like a paperclip, you have to move the picture frame closer to model space to make it appear larger. For example: 8 = 1   .

ADJUSTING THE VIEWPORT SCALE

The following will take you through the process of adjusting the scale within a viewport.

1. **Open** a drawing.

2. Select a **Layout** tab. (paper space)

3. Cut a new Viewport or unlock an existing Viewport.

4. **Zoom / All** to display all of the drawing limits.

5. Adjust the scale
   A. You must be in Paper Space.
   B. Select the Viewport Frame.
   C. Unlock Viewport, if locked.
   D. Select the Viewport Scale down arrow.
   E. Select the scale from the list of scales.

Do not select this lock tool

6. Lock the Viewport

Note: If you would like to add a scale that is not on the list:
   1. Type **Scalelistedit**
   2. Select **Add** button.
   3. Enter Scale name to display in scale list.
   4. Enter Paper and drawing units.
   5. Select the **OK** button.

ANNOTATIVE OBJECTS

In lesson 26 you created a new text style and dimension style. You also added the **"Annotative" property** to both simply by placing a check mark in the Annotative box.

The Text style and Dimension Style shown above are now **Annotative**.
Notice the Annotative symbol ⚘ beside the Name. (No symbol by Standard name.)

"Annotative objects" are scaled automatically to match the scale of the viewport.
For example, if you want the text, inside the viewport, to print .200 in height and the viewport scale is 1:2 AutoCAD will automatically scale the text height to .400. The text height needs to be scaled by a factor of 2 to compensate for the model space contents appearing smaller.

The easiest way to understand how **Annotative property** works is to do it.
So try this example. You will need this file for Lesson 28 also.

1. Open **My Decimal Setup**

2. Select the **Model** tab.
3. Draw a rectangle   3.00 Lg   X   1.50 wide   Use layer Object line.

4. Select the **A Size** tab.

Continued on the next page...

ANNOTATIVE OBJECTS....continued

5. Erase the existing single viewport frame.

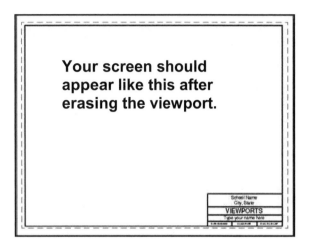

6. Cut 2 new viewports as shown below. Use Layer Viewport

7. Activate the left viewport (double click inside the viewport frame) and do the following:
   A. Adjust the scale of the viewport to **1 : 1** (Refer to page 27-3)
   B: PAN to place rectangle in the center of the viewport. (Refer to page 26-10)
   C. Lock the viewport (Refer to page 27-3)

8. Activate the right viewport (click inside the viewport frame) and do the following:
   A. Adjust the scale of the viewport to **1 : 4** (Refer to page 27-3)
   B: PAN to place rectangle in the center of the viewport. (Refer to page 26-10)
   C. Lock the viewport   (Refer to page 27-3)

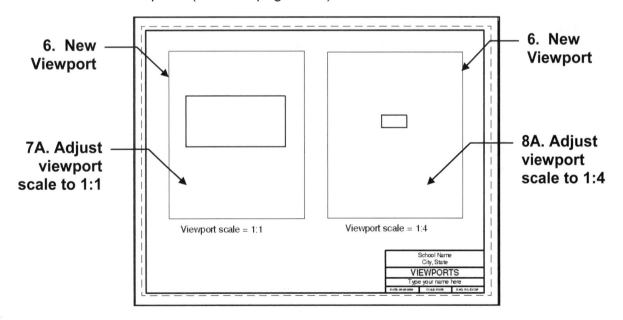

Your screen should appear approximately like this.

Continued on the next page...

ANNOTATIVE OBJECTS....continued

9. Activate the left hand viewport. (Double click inside left hand viewport)

10. Add .200 height text **Scale 1:1** as shown using text style "Text-Classic".
    (Note: Text style "Text-Classic" is annotative)

11. Add the dimension shown using dimension style "Dim-Decimal".
    (Note: Dimension style "Dim-Decimal" is annotative)

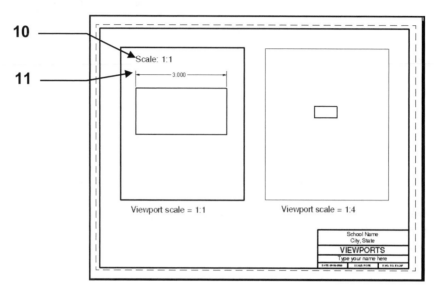

12. Activate the <u>right hand</u> viewport. (Click inside right hand viewport)

13. Add .200 height text **Scale 1:4** as shown using text style "Text-Classic".
    (Note: Text style "Text-Classic" is annotative)

14. Add the dimension shown using dimension style "Dim-Decimal".
    (Note: Dimension style "Dim-Decimal" is annotative)

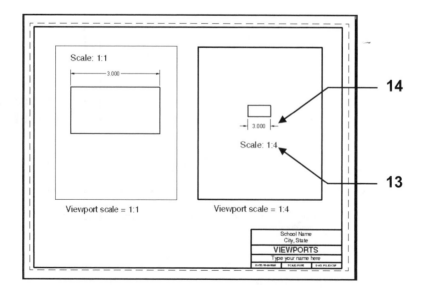

Continued on the next page...

ANNOTATIVE OBJECTS....continued

Notice that the text and dimensions appear the <u>same size in both viewports</u>.

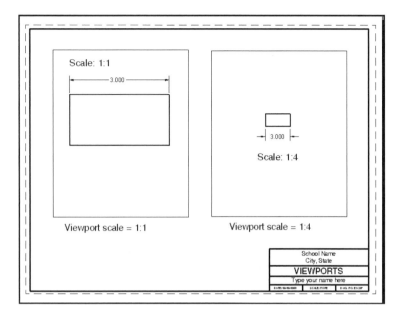

15. Now select the **Model** tab.

Notice there is 2 sets of text and dimensions.

One set has the annotative scale of 1:1 and will be visible only in a 1:1 viewport.
One set has the annotative scale of 1:4 and will be visible only in a 1:4 viewport.
But you see both sets when you select the model tab.

(In Lesson 28 you will learn how to assign multiple annotative scales to one set of text or dimensions so you will not have duplicate sets in the model tab.)

Continued on the next page...

ANNOTATIVE OBJECTS....continued

16. Place your cursor on any of the text or dimensions. An "Annotative symbol" ⚠ will appear. This indicates that this object is annotative and it has only one annotative scale. (In lesson 28 you will learn how to assign multiple annotative scales to a single annotative object.)

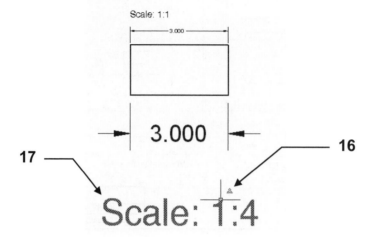

17. Click once on the **Scale: 1:4** text. The Quick properties box should appear.

Notice the text height is listed twice.

Paper text height = .200 This is the height that you selected when placing the text. When the drawing is printed the text will print .200.

Model text height = .800 This is the desired height of the text (.200) factored by the viewport scale (1:4). The viewport scale is a factor of 4. (4 X .200 = .800)

18. **IMPORTANT:** Save this drawing as: **Annotative objects**
    (You will need it for Lesson 28)

Summary:
If an object is annotative AutoCAD automatically adjusts the scale of the object to the viewport scale. The most commonly used Annotative Objects are: Dimensions, Text, Hatch and Multileaders. Refer to the Help menu for more.
In lesson 28 you will learn how to assign multiple annotative scales to a single annotative object.)

PAPER SPACE DIMENSIONING

Some people prefer to dimension in paper space. Paper space dimensions are Trans-spatial. Trans-spatial means that you may place the dimension in paper space while the object you are dimensioning is in model space. Even though the dimension is in paper space it is actually attached to the object in model space.

For example,
1. You draw a house in model space.
2. Now select the layout tab.
3. Cut a viewport so you can see the drawing of the house.
4. Go to model space, adjust the scale of the viewport (model space) and lock it.
5. Now go to paper space and dimension the house.

SHOULD YOU DIMENSION IN PAPER SPACE OR MODEL SPACE?

This is your choice. Personally, I dimension in either space depending on the situation. Here are some things to consider.

Paper Space dimensioning

Pro's
1. You never have to worry about the viewport scale.
2. All dimensions will have the same appearance in all viewports.
3. If you Xref the drawing the dimensions do not come with it.
   (Xref is not discussed in this workbook)

Con's
1. Qdim is not trans-spatial. (You can only use it in model space.)
2. If you move the objects in model space, or adjust the viewport scale, sometimes the dimensions do not move with them. (If this happens type Dimregen <enter>)
3. Dimensions will not appear in the Model tab. They appear only in the layout tab in which they were placed. If you go to another layout tab they will not appear.

Annotative dimensioning in Model Space

Pro's
1. Dimensions always appear and move with the objects.
2. Dimensions will appear in all layout tabs within viewports.

Con's
1. If you Xref the drawing the dimensions come also.

EXERCISE 27A

CREATE A MASTER FEET-INCHES SETUP TEMPLATE

The following instructions will guide you through creating a "Master" Feet-inches setup template. The "2010-Workbook Helper" is an example of a Master setup template. Even though the screen appears blank, the actual file is full of settings, such as: Units, Drawing Limits, Snap and Grid settings, Layers, Text styles and Dimension Styles. Once you have created this "Master" template, you just open it and draw. No more repetitive inputting of settings.

NEW SETTINGS

A. Begin your drawing with a different template as follows:
    1. Select the **NEW** command.
    2. Select template file **acad.dwt** from the list of templates. (**Not** acad3D.dwt)
    3. Select **Open**

B. Set the drawing **Units** as follows:
    1. Type **UNITS <enter>**
    2. Change the **Type** and **Precision** as shown
    3. Select **OK** button

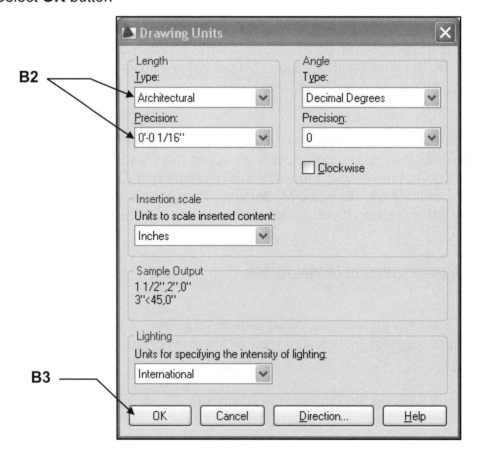

Continued on the next page...

EXERCISE 27A continued

C. Set the **Drawing Limits** (Size of the drawing area) as follows:
1. Type **Limits <enter>**
2. Enter Lower left corner = 0.000, 0.000
3. Enter Upper right corner = **44' , 34'** (Notice this is feet, not inches)
4. Use **" ZOOM / ALL"** to view the new limits
5. Set your **Grids** to **ON** to display the limits (drawing area).

D. Set the Grids and Snap as follows:
1. Type **DS<enter>**
2. Select the **Snap and Grid** tab
3. Change the settings as shown below.
4. Select the **OK** button.

D2

Important: Uncheck Adaptive Grid

D4

Continued on the next page...

EXERCISE 27A continued

E. Change **Lineweight** settings as follows:
1. Right click on Lineweight button on status line
2. Select **Settings**
3. Change to inches and adjust the Display scale as shown below.
4. Select the **OK** button.

F. Create **new Layers** as follows:
1. Load the linetype **Dashed**. (Refer to page 3-16 for instructions)

2. Assign names, colors, linetypes, lineweights and plotability as shown below:
    (Refer to 3-15 for instructions)

Notice
Lineweights

S..	Name	On	Freeze	Lock	Color	Linetype	Lineweight	Plot Style	Plot
	0				white	Continuous	Default	Color_7	
	BORDER				white	Continuous	0.039"	Color_7	
	CABINETS				cyan	Continuous	0.016"	Color_4	
	Defpoints				white	Continuous	Default	Color_7	
	DIMENSION				blue	Continuous	Default	Color_5	
	DOORS				green	Continuous	0.016"	Color_3	
	ELECTRICAL				cyan	Continuous	Default	Color_4	
	FURNITURE				magenta	Continuous	0.016"	Color_6	
	HARDSCAPE				white	Continuous	0.024"	Color_7	
	HATCH				green	Continuous	Default	Color_3	
	LANDSCAPE				green	Continuous	0.016"	Color_3	
	MISC				cyan	Continuous	Default	Color_4	
	PLANTS				green	Continuous	0.016"	Color_3	
	PLUMBING				9	Continuous	Default	Color_9	
	ROOF				white	Continuous	0.016"	Color_7	
	TEXT				blue	Continuous	Default	Color_5	
	VIEWPORT				green	Continuous	Default	Color_3	
✓	WALLS				red	Continuous	0.031"	Color_1	
	WINDOWS				green	Continuous	0.016"	Color_3	
	WIRING				cyan	DASHED	Default	Color_4	

Continued on the next page...

EXERCISE 27A continued

G. Create 2 new **Text Styles** as follows:
1. Select **Text Style** (Refer to page 25-2)
2. Create the text style **Text-Classic** and **Text-Arch** using the settings shown below.

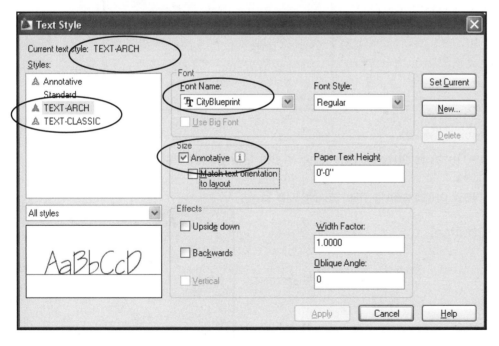

THIS NEXT STEP IS VERY IMPORTANT

H. **Save** all the settings you just created as follows:
1. Save as: **My Feet-Inches Setup**

I. Now continue on to **Exercise 27B**...you are not done yet.

EXERCISE 27B

PAGE SETUP

Now you will select the printer and the paper size to use for printing.
You will use the Layout1 tab (Paper space).

A. Open **My Feet-Inches Setup** (If not already open)

B. Select **Layout1** tab.

Refer to page 26-3 if you do not have these tabs.

Note: If the "Page Setup Manager" dialog box shown below does not appear automatically, right click on the Layout tab and select Page Setup Manager.

C. Select the **New...** button.

D. Select the **<Default output device>** in the Start with: list.

E. Enter the New page setup name: **Setup B**

F. Select **OK** button.

(I am assuming that your computer is attached to a printer. If not select Layout1)

Continued on the next page...

EXERCISE 27B continued

This is where you will select the **print device**, **paper size** and the **plot offset**.

G. Select the **Printer / Plotter**
   Note: Your current system printer should already be displayed here. If you prefer another select the down arrow and select from the list. If the preferred printer is not in the list you must configure the printer. Refer to Appendix-A for instructions.)

H. Select the **Paper Size**

Notice the name you entered is now displayed as the page setup name.

I. Select **Plot Offset**

G
(Yours will be different)

H
(Yours may be 8-1/2 x 11)

I
(Should stay at 0.00000)

J. Select **OK** button.

J

K. Select **Setup B**.

L. Select the **Set Current** button.

M. Select the **Close** button.

K

L

M

Continued on the next page...

EXERCISE 27B continued

You should now have a sheet of paper displayed on the screen.
This sheet is the size you specified in the "Page Setup".
This sheet is in front of Model Space.

The dashed line represents the maximum printing area for the printing device that you selected.
Any object outside of this area will not print.

N

New layout
From template...
Delete
Rename
Move or Copy...
Select All Layouts

Activate Previous Layout
Activate Model Tab

Page Setup Manager...
Plot...

Import Layout as Sheet...
Export Layout to Model...

Hide Layout and Model tabs

Model \ Layout1 / Layout2 /

Rename the Layout tab

N. Right click on the Layout1 tab and select **Rename** from the list.

O. Enter the New Layout name **A Size**   Model \ A Size / Layout2 /

P. ***Very important:*** Save as **My Feet-Inches Setup** again.

Q. Continue on to **Exercise 27C**.....you are not done yet.

EXERCISE 27C

CREATE A BORDER AND TITLE BLOCK

A. Draw the <u>Border rectangle</u> as large as you can within the dashed lines approximately as shown below using <u>Layer Border</u>.

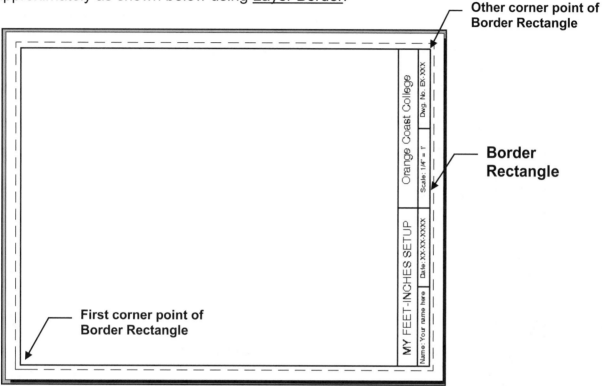

B. Draw the Title Block as shown using:
   1. Layers <u>Border</u> and <u>Text</u>.
   2. **Single Line Text** ; Justify **<u>Middle</u>** in each rectangular area.
      (Hint: Draw diagonal lines to find the middle of each area, as shown)
   3. Text Style= Text-Classic   Text height varies.

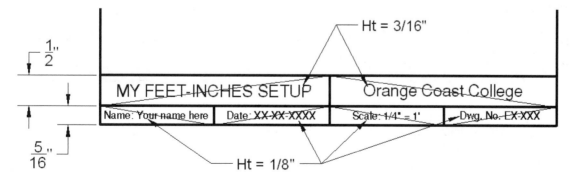

C. *Very important:* Save as **My Feet-Inches Setup** again.

D. Continue on to **Exercise 27D**.....you are not done yet.

EXERCISE 27D

CREATE A VIEWPORT

The following instructions will guide you through creating a VIEWPORT in the Border Layout sheet. Creating a viewport has the same effect as cutting a hole in the sheet of paper. You will be able to see through the viewport frame (hole) to Modelspace.

A. Open **My Feet-Inches Setup** (If not already open)
B. Select the **A Size** tab.
C. Select layer **Viewport**
D. Type: **MV <enter>**
E. Draw a Single viewport approximately as shown. (Turn off OSNAP temporarily)

F. Adjust the Viewport Scale as follows: **F1**
    F1. Confirm **Model button** showing. MODEL
    F2. Select **Zoom / All**
    F3. Select the **Viewport Scale** of **1/4" = 1'-0"** (Note: inches = feet)

G

F3

G. **Lock** the Viewport (Refer to page 27-3)

H. ***Very important....*** Save as **My Feet-Inches Setup** again.

I. Continue on to **Exercise 27E**....you are not done yet.

EXERCISE 27E

PLOTTING FROM THE LAYOUT

The following instructions will guide you through the final steps for setting up the master template for plotting. These settings will stay with **My Feet-Inches Setup** and you will be able to use it over and over again.

In this exercise you will take a short cut by **importing** the **Plot Setup A** from **My Decimal Setup**. (Note: If the Printer, Paper size and Plot Scale is the same you can use the same Page Setup.)

Note: If you prefer not to use Import, you may go to 26-33 and follow the instructions for creating the Plot- page setup.

A. Open **My Feet-Inches Setup** (If not already open)

B. Select the **A Size** layout tab.

   You should be seeing your Border and Title Block now.

C. Select **Plot** command.

D. Select **Import...** from the Page Setup drop down menu

E. Find **My Decimal Setup.dwt** as follows:
   E1. Select Files of type: **Template [.dwt]**
   E2. Select **My Decimal Setup.dwt**
   E3. Select **OK** button.

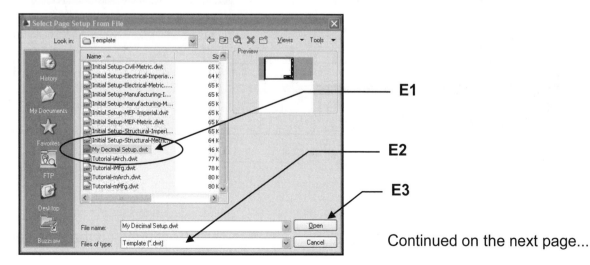

Continued on the next page...

EXERCISE 27E continued

F. Select **Plot Setup A** from the Page Setups list.

G. Select the **OK** button.

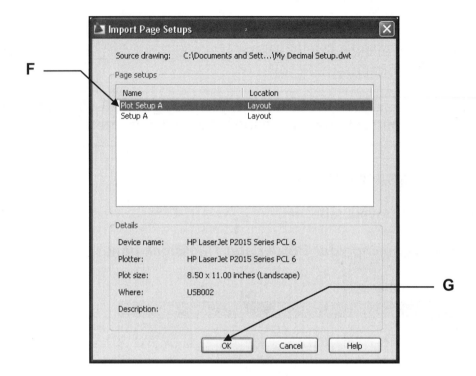

H. Select **Plot Setup A** from the Page Setup drop down list.
   (Note: It was not there before. You just imported it in from My Decimal Setup)

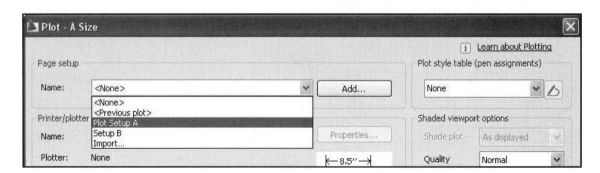

Continued on the next page...

EXERCISE 27E continued

I. Check all the setting.

J. Select **Preview** button.

   If the drawing appears as you would like it, press the **Esc** key and continue

   If the drawing does not look correct, press the **Esc** key and check all your
      settings, then preview again.

Note: The Viewport will not appear in the Preview because it is on a no plot layer.

K. Select the **Apply to Layout** button.
   (The settings are now saved to the layout tab for future use.)

L. If your computer **is** connected to the plotter / printer selected, select the **OK** button
   to plot, then proceed to **N**.

M. If your computer is **not** connected to the plotter / printer selected, select the Cancel
   button to close the Plot dialog box and proceed to **N**.

N. ***Very important....*** Save as **My Feet-Inches Setup** again.

O. Continue on to **Exercise 27F**....you are not done yet.

EXERCISE 27F

CREATE A NEW DIMENSION STYLE

A. Open **My Feet-Inches Setup** (If not already open)

B. Select the **Dimension Style Manager** command (Refer to 16-9)

C. Select the **NEW** button.

D. Enter **DIM-ARCH** in the "New Style Name" box.

E. Select **STANDARD** in the **"Start With:"** box.

F. Select **Annotative** box.

G. Select the **CONTINUE** button.

EXERCISE 27F continued

H. Select the **Primary Units** tab and change your settings to match the settings shown below.

H

Do not select the OK button yet

EXERCISE 27F continued

I. Select the **Lines** tab and change your settings to match the settings shown below.

I

Do not select the OK button yet ⎯

EXERCISE 27F continued

J. Select the **Symbols and Arrows** tab and change your settings to match the settings shown below.

J

Do not select the OK button yet

EXERCISE 27F continued

K. Select the **Text** tab and change your settings to match the settings shown below.

K

Do not select the OK button yet

EXERCISE 27F continued

L. Select the **Fit** tab and change your settings to match the settings shown below.

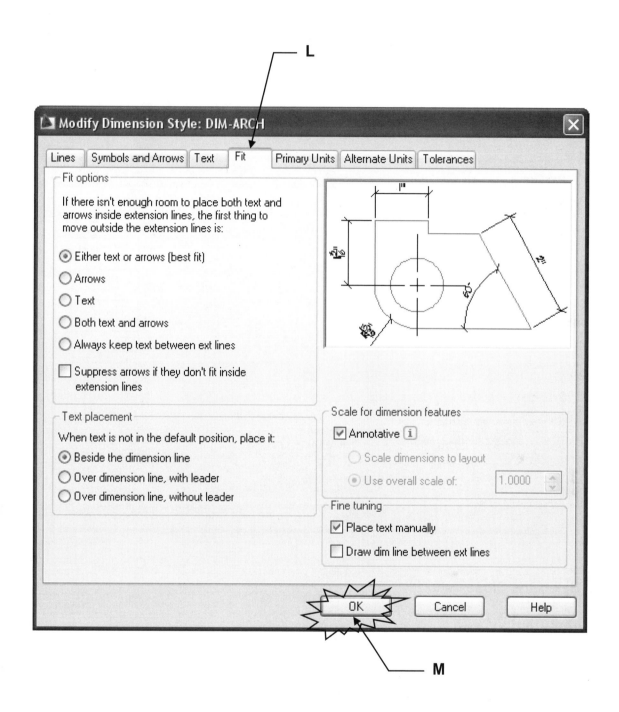

M. Now select the **OK** button.

EXERCISE 27F   continued

Your new **DIM-ARCH** dimension style should now be in the list.

N. Select the **Set Current** button to make your new style **DIM-ARCH** the style that will be used.

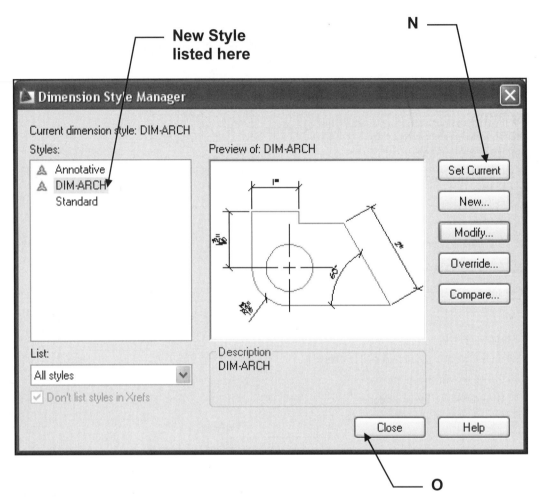

New Style listed here

N

O

O. Select the **Close** button.

P. **Important:**
Save your drawing as **My Feet-Inches Setup** again.

Q. You are almost done. Now Save all of this work as a **Template. (Refer to 2-3)**
   a. Select **Application Menu / Save As / AutoCAD Drawing Template**
   b. Type: **My Feet-Inches Setup**
   c. Select **Save** button.

Again, I know that seemed like a lot of work but you are now ready to use this master setup template to create and plot many drawings in the future.

EXERCISE 27G

INSTRUCTIONS:
1. Select **New** and select **My Feet-Inches Setup.dwt** (template)
2. Draw the classroom shown below.
3. Use Layers Walls, Furniture, Doors and Windows.
4. Dimension as shown inside the viewport.
5. Use Dimension Style "Dim-Arch" and Layer: Dimension
6. Edit the Title and Ex-XX by double clicking on the text. Do not erase and replace.
7. Save as **EX27G**
8. Plot using Page Setup **Plot Setup A** (Refer to page 27-21)

Drawing hints:
1. Consider using **Array** (Lesson 13) and **Offset** (Lesson 12).
2. Remember you have to Unlock the Viewport to use Pan inside the Viewport.
3. If you change the scale of the Viewport, change the viewport scale back to 1/4"=1'
   and lock the viewport.
4. If your dimensions are gigantic, you are not inside the viewport.

EXERCISE 27H

INSTRUCTIONS:
1. Select **New** and select **My Decimal Setup.dwt** (template)
2. <u>Unlock</u> the Viewport.
3. Change the Viewport Scale to 8=1
4. Lock the Viewport
5. Turn **Snap** off or change your grid and snap relative to the new viewport scale.
6. Draw the Paperclip on layer Object line.
7. Dimension as shown using Dimension Style: Dim-Decimal
8. Select the Viewport layer and cut a second small viewport next to the Title Block
9. Use **Zoom / All** inside of the **small** viewport. (You should now see the paper clip)
10. Change the Viewport Scale of the **small** Viewport to **1 = 1** and <u>Lock.</u>
11. Add "**Actual Size**" text .200 ht in Paper Space   Use: layer Text.
12. Edit the Title and Ex-XX by double clicking on the text. Do not erase and replace.
13. Save as **EX27H**
14. Plot using Page Setup **Plot Setup A** (Refer to page 26-34)

Drawing hints: The entire Paper clip can easily be drawn using Offset (Lesson 12) and Fillet (Lesson 7) or Circles and trim.

27-30

LEARNING OBJECTIVES

After completing this lesson, you will be able to:

1. Assign multiple annotative scales to a single annotative object
2. Assigning annotative scales to Hatch sets

LESSON 28

ASSIGNING MULTIPLE ANNOTATIVE SCALES

In the previous lesson you learned how annotative text and dimensions are automatically scaled to the viewport scale. But in order to have an annotative object appear in both viewports you placed 2 sets of text and dimensions. In this lesson you will learn how to easily assign multiple annotative scales to a single text string or dimension so you need not duplicate them each time you create a new viewport. You will just assign an additional annotative scale to the annotative object.

Again, the easiest way to understand this process it to do it. The following is a step by step example.

1. Open **Annotative Objects.dwg** from the previous lesson. *If you did not complete the example from the previous lesson go back and do it now. (Refer to page 27-4)*

2. Make the right hand viewport active. (Double click inside the viewport)

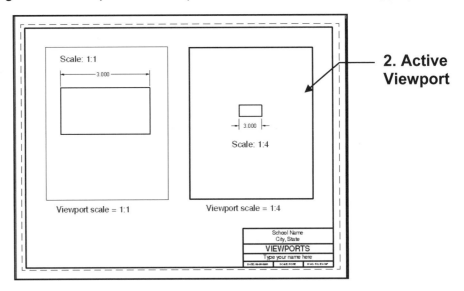

3. Erase the text and the dimension in the right hand viewport only.

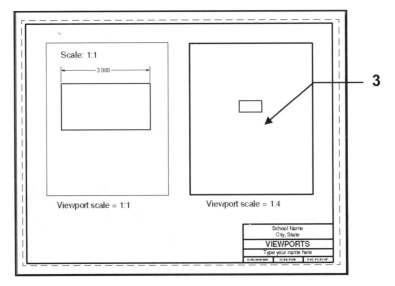

Continued on the next page...

ASSIGNING MULTIPLE ANNOTATIVE SCALES continued

4. Display all annotative objects in all viewports as follows:
A. Select the **Annotation Visibility** button located in the lower right corner of the drawing status bar.

4A. Annotation Visibility button

Light bulb should be yellow

 ON (<u>Yellow</u> light bulb) Displays **all** Annotation objects in **all** viewports. (example below)

 OFF (<u>Blue</u> light bulb)   Displays **only** Annotation objects that have an annotative scale that matches the viewport scale.

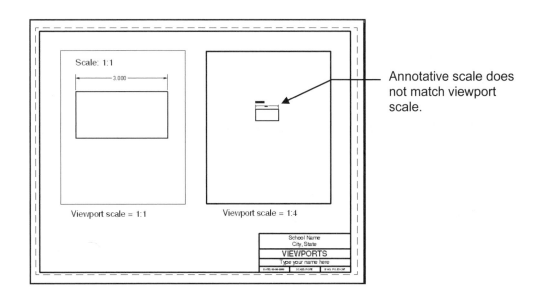

Annotative scale does not match viewport scale.

The dimensions and text are now displayed in both viewports. But the annotative scale of the dimensions and text in the right hand viewport do not match the scale of the viewport. (Notice they are smaller) **The scale of annotative objects must match the scale of the viewport.** *Follow the steps on the next page to assign multiple annotative scales to an annotative object.*

Continued on the next page...

ASSIGNING MULTIPLE ANNOTATIVE SCALES continued

5. Place your cursor near the **dimension** in the right hand viewport. Notice the <u>single</u> **Annotation symbol.** This single symbol indicates the annotative dimension has only one annotative scale assigned to it.

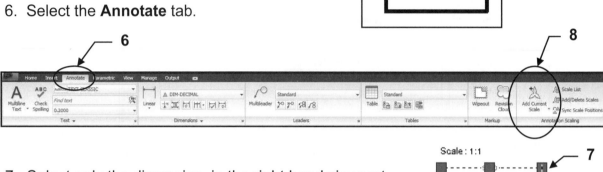

6. Select the **Annotate** tab.

7. Select <u>only</u> the <u>dimension</u> in the <u>right hand viewport</u>.

8. Select the **Add Current Scale** tool on the <u>Annotation Scaling</u> panel.

The annotative dimension should have increased in size as shown below.

9. <u>Turn OFF</u> the **Annotation Visibility.** (Click on button. The light should turn blue)

9. Annotation Visibility button should turn blue

Continued on the next page...

ASSIGNING MULTIPLE ANNOTATIVE SCALES continued

*Notice the text in the right hand viewport is no longer visible. When the **Annotative Visibility** is OFF only the annotative objects that match the viewport scale will remain visible. The dimension is the only annotative object that has the 1:4 annotative scale assigned to it.*

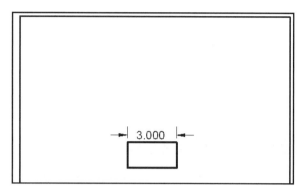

Annotative Visibility OFF

10. Place your cursor near the dimension. Notice 2 annotation symbols appear now. This indicates 2 annotation scales have been assigned to the annotative object.

10

11. Click on the dimension to display the grips and drag the dimension away from rectangle approximately as shown below. (The dimension was too close to the rectangle) Notice the dimension in the left hand viewport did not move. They can be moved individually.

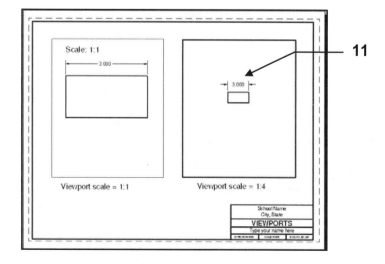

11

HOW TO REMOVE AN ANNOTATIVE SCALE

If you have an annotative object such as a dimension, that you would like to remove from a viewport, you must remove the annotative scale that matches the viewport scale. **Do not delete** the dimension because it will also be deleted from all of the other viewports. This sounds complicated but is very easy to accomplish.

Problem: I would like to remove **dimension A** from the right hand viewport but I do not want to remove **dimension B** in the lower viewport.

Solution: I must remove the 1/4" = 1" annotation scale from **dimension A**.

(Step by step instructions below.)

Step 1.

1. Select the **Annotate** tab.

2. Select **dimension A** shown above. (You must be inside the viewport)

3. Select **Add / Delete Scales** tool located on the **Annotation Scaling** panel.

Continued on the next page...

HOW TO REMOVE AN ANNOTATIVE SCALE continued

4. Select the annotation scale to remove. (1/4" = 1')

5. Select the **Delete** button.
   (Remember, you are deleting the annotative scale not the actual dimension. The dimension still exists but it will not have an annotative scale of 1/4"=1' assigned to it. As a result it will not be visible within any viewport that has been scaled to 1/4" = 1')

6. Select the **OK** button.

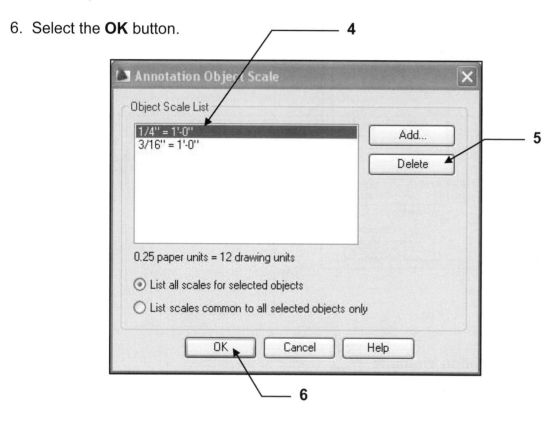

Note: Dimension A has been removed and **dimension B** remains.

Dimension A removed from viewport

Dimension B remains

ANNOTATIVE HATCH

When you add cross-hatching to a drawing it is usually in just one area or viewport. But if you would like it shown in all viewports follow the steps below.

1. Draw the hatch in one of the viewports using **Annotative** hatch.
   a. Select the **Annotative** box

2. Turn ON **Annotation Visibility** (Refer to page 28-3)

3. Select the <u>Hatch Set</u> to change.

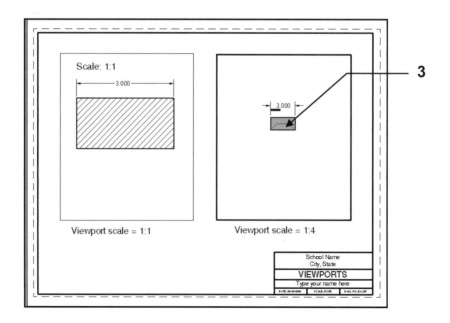

ANNOTATIVE HATCH

4. Select the "**Add Current Scale**" tool located on the **Annotate** tab, **Annotation Scaling** panel.

5. Turn OFF **Annotation Visibility**

Now the appearance of the hatch sets should be identical in both viewports.

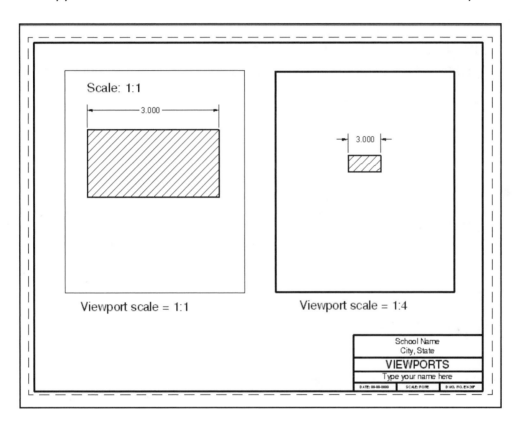

EXERCISE 28A

The following exercise will give you practice adding Annotative Object Scales to existing dimensions in order to display them within viewports with different scales. Follow the instructions below as you manipulate the drawing to appear as shown on page 28-13.

STEP 1

1. Open **EX-27G**

2. Select the **A size** tab.

3. **Erase** the 1 existing Viewport
   (The classroom will disappear from paperspace because the hole is gone.)

4. Select the Viewport Layer and **Create** 4 new viewports approximately as shown below.

Notice each viewport is a hole in paperspace
and you are looking through to modelspace

EXERCISE 28A CONTINUED

STEP 2

5. **Adjust the scale** of each Viewport as indicated.

6. Use **Pan** to move the image within the viewport without changing the scale.

7. Add the **VP Scale** labels.
   Place them in Paper space. Text style: Text-Classic   1/8" ht.

8. **Lock** each viewports.

Notice the annotative dimensions appear only in the 1/4" = 1' viewport.
The dimensions currently have only 1/4"=1' annotation scale.

EXERCISE 28A CONTINUED

Step 3

1. Turn on **Annotation Visibility** (<u>yellow</u> light bulb   Refer to 28-3)

You should see dimensions in all viewports now.

EXERCISE 28A CONTINUED

Step 4

1. Add **Annotative Object Scales** to the dimensions shown in the viewports below:

   A. Select a Viewport (double click inside viewport)

   B. Select the dimensions that you want to stay visible in that viewport.

   C. Select the **Annotate** panel. (Refer to 28-4)

   D. Select   **Add Current Scale** button
   This will automatically add the current viewport scale to the selected objects.

   E. Go on to the next Viewport and repeat A thru D above until they are all done.

Step 5

1. Turn OFF **Annotative Visibility**   (blue light bulb)
   The unwanted dimensions should have disappeared.

2. Save as: **EX28A**

EXERCISE 28B

The following exercise will give you practice with Hatch within scaled viewports using **Annotative Object Scale** option. Follow the instructions below while you manipulate the drawing to appear as shown.

STEP 1

1. Open **My Decimal Setup**

2. **Erase** the 1 <u>existing Viewport</u>.

3. <u>Select the Viewport</u> Layer and **Create 3 new viewports** approximately as shown.

4. **Adjust the scale** of each Viewport as indicated.

5. **Lock** each viewport.

6. Activate the 1 = 1 viewport. (Double click inside the viewport)

7. **Draw** a 2" X 2" Rectangle inside the 1 = 1 Viewport as shown using <u>Layer Object</u>. (The rectangle will appear in the other 2 Viewports because all viewports display modelspace.)

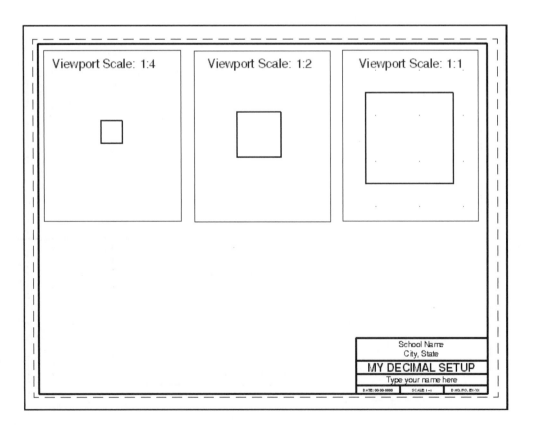

Continued on the next page...

EXERCISE 28B CONTINUED

STEP 2

1. Select the layer **Hatch**

2. **Hatch** the 1 = 1 Rectangle using:
   A. Pattern **Ansi 31**
   B. Angle **0**
   C. Scale **1.000**
   D. Select the **Annotative** box.

Note:
If a Hatch Set appears in all Rectangles you probably forgot to select the Annotative box <u>or</u> you have Annotation visibility ON.

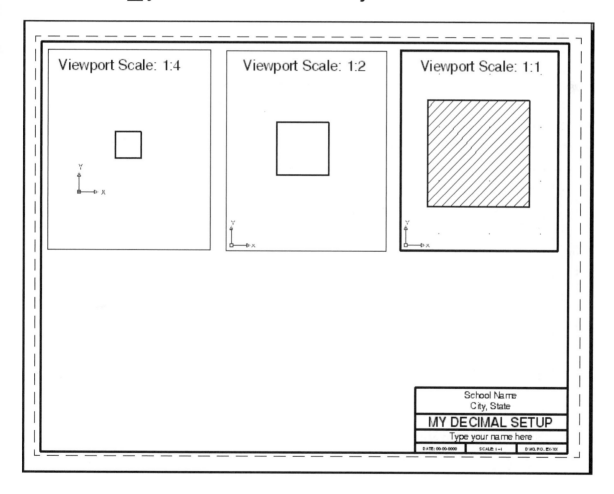

Continued on the next page...

EXERCISE 28B CONTINUED

STEP 3

1. Turn ON **Annotation Visibility** 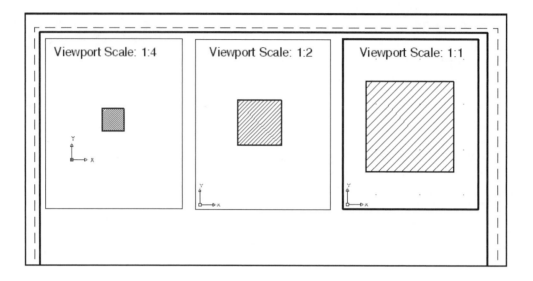 (<u>yellow</u> light bulb   Refer to 28-3)

2. Assign the correct Annotative scale to the hatch sets in viewport 1:4 and 1:2.

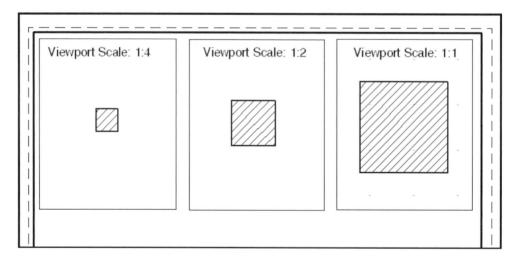

3. Turn ON **Annotation Visibility** (<u>blue</u> light bulb   Refer to 28-3)

4. Save as: **EX28B**

LEARNING OBJECTIVES

After completing this lesson, you will be able to:

1. Understand Blocks
2. Create a Block
3. Insert a Block
4. Re-define and Purge unused Blocks
5. Create Mulitleaders with blocks attached
6. Use the Collect multileader tool.

LESSON 29

BLOCKS

A **BLOCK** is a group of objects that have been converted into ONE object. A Symbol, such as a transistor, bathroom fixture, window, screw or tree, is a typical application for the block command. First a BLOCK must be created. Then it can be INSERTED into the drawing. An inserted Block uses less file space than a set of objects copied.

CREATING A BLOCK

1. First draw the objects that will be converted into a Block.

   For this example a circle and 2 lines are drawn.

2. Select the **BLOCK MAKE** command using one of the following:

   Ribbon = Insert tab / Block panel /

   Keyboard = B <enter>

3. Enter the New Block name in the **Name** box.

BLOCKS....continued

4. Select the **Pick Point** button. (Or you may type the X, Y and Z coordinates.)
    The Block Definition box will disappear and you will return temporarily to the drawing.

5. Select the location where you would like the insertion point for the Block.
    Later when you insert this block, the block will appear on the screen attached to the cursor at this insertion point. Usually this point is the CENTER, MIDPOINT or ENDPOINT of an object.

Notice the coordinates for the base point are now displayed. (Don't worry about this. Use Pick Point and you will be fine)

6. Select an option.

    It is important that you select one and understand the options below.

    Retain
    If this option is selected, the original objects will stay visible on the screen after the block has been created.

    Convert to block
    If this option is selected, the original objects will disappear after the block has been created, but will immediately reappear as a block. It happens so fast you won't even notice the original objects disappeared.

    Delete
    If this option is selected, the original objects will disappear from the screen after the block has been created. (This is the one I use most of the time)

7. Select the **Select Objects** button.

    The Block Definition box will disappear and you will return temporarily to the drawing.

8. Select the objects you want in the block, then press <enter>.

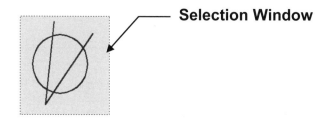

Selection Window

BLOCKS....continued

The Block Definition box will reappear and the objects you selected should be illustrated in the Preview Icon area.

10 ⎯⎯   ⎯⎯ **Preview**

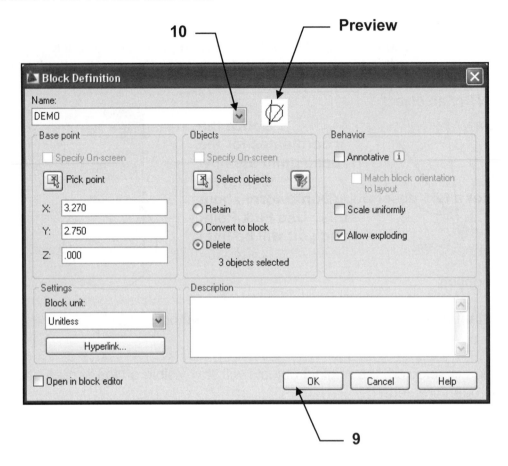

9. Select the **OK** button.
   The new block is now stored in the drawing's block definition table.

10. To verify the creation of this Block, select [⬚] **Create** again, and select the Name (▾). A list of all the blocks, in this drawing, will appear.
    (Refer to page 29-6 for inserting instructions)

BLOCKS....continued

ADDITIONAL DEFINITIONS OF OPTIONS

Block Units
You may define the units of measurement for the block. This option is used with the "Design Center" to drag and drop with Autoscaling. The Design Center is an advanced option and is not discussed in this book.

Hyperlink
Opens the **insert Hyperlink dialog box** which you can use to associate a hyperlink with the block.

Description
You may enter a text description of the block.

Scale Uniformly
Specifies whether or not the block is prevented from being scaled non-uniformly during insertion.

Allow Exploding
Specifies whether or not the block can be exploded after insertion.

HOW LAYERS AFFECT BLOCKS

If a block is created on Layer 0:

1. When the block is inserted, it will take on the properties of the current layer.
2. The inserted block will reside on the layer that was current at the time of insertion.
3. If you Freeze or turn Off the layer the block was inserted onto, the block will disappear.
4. If the Block is **Exploded**, the objects included in the block will revert to their original properties of layer 0.

If a block is created on Specific layers:

1. When the block is inserted, it will retain its own properties. It **will not** take on the properties of the current layer.
2. The inserted block **will reside** on the current layer at the time of insertion.
3. If you **freeze** the layer that was current at the time of insertion the block <u>will</u> disappear.
4. If you turn **off** the layer that was current at the time of insertion the block <u>will not</u> disappear.
5. If you **freeze** or turn **off** the blocks original layers the block <u>will</u> disappear.
6. If the Block is **Exploded**, the objects included in the block will go back to their original layer.

INSERTING BLOCKS

A **BLOCK** can be inserted at any location within the drawing. When inserting a Block you can **SCALE** or **ROTATE** it.

1. Select the **INSERT** command using one of the following:

   Ribbon = Insert tab / Block panel /

   Keyboard = Insert <enter>

2. Select the **BLOCK** name.
   a. If the block is already in the drawing that is open on the screen, you may select the block from the drop down list shown above
   b. If you want to insert an entire drawing, select the Browse button to locate the drawing file.

3. Select the **OK** button.

   This returns you to the drawing and the selected block should be attached to the cursor.

4. Select the insertion location for the block by moving the cursor and pressing the left mouse button or typing coordinates.

   Command: _insert
   Specify insertion point or **[Basepoint/Scale/X/Y/Z/Rotate]:**

   NOTE: If you want to change the **basepoint, scale** or
   rotate the block before you actually place the block,
   press the right hand mouse button and you may
   select an option from the menu or select an option
   from the command line menu shown above.

   You may also "**preset**" the insertion point, scale or
   rotation. This is discussed on the next page.

INSERTING BLOCKS....continued

PRESETTING THE <u>INSERTION POINT</u>, <u>SCALE</u> or <u>ROTATION</u>

You may preset the **Insertion point, Scale or Rotation** in the <u>**INSERT**</u> box instead of at the command line.

1. Remove the check mark from any of the **"Specify On-screen"** boxes.
2. Fill in the appropriate information describe below:

Insertion point
Type the X and Y coordinates <u>from the Origin</u>. The Z is for 3D only.
The example below indicates the block's insertion location will be 5 inches in the X direction and 3 inches in the Y direction, <u>from the Origin</u>.

Scale
You may scale the block proportionately by typing the scale factor in the X box and then check the <u>Uniform Scale box</u>. If you selected "Scale uniformly" box when creating the block this option is unnecessary and not available.
If the block is to be scaled non-proportionately, type the different scale factors in both X and Y boxes.
The example below indicates that the block will be scale proportionate at a factor of 2.

Rotation
Type the desired rotation angle relative to its current rotation angle.
The example below indicates the block will be rotated 45 degrees from its originally created angle orientation.

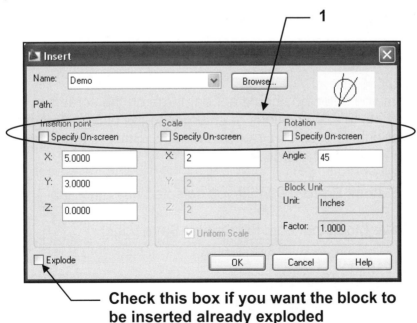

Check this box if you want the block to be inserted already exploded

RE-DEFINING A BLOCK

How to change the design of a block previously inserted.

1. Select **Block Editor** using one of the following:

   Ribbon = Insert tab / Block panel /
   Keyboard = bedit

   *The "**Edit Block Definition**" dialog box will appear.*

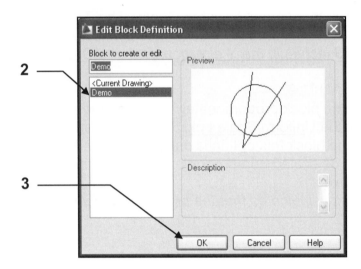

2. Select the name of the Block that you wish to change.

3. Select the **OK** button.

4. The Block that you selected should appear large on the screen.
   You may now make any additions or changes to the block. You can change tabs and use other panels such as Draw and Modify. But you must return to the **Block Editor** tab to complete the process.

5. Return to the **Block Editor** button if you selected any other tab.

6. Select the **Save block** tool from the **Open / Save** panel.

7. Select the **Close** tool.

You will be returned to the drawing and **ALL** of the blocks with the same name will be updated with the changes that you made.

PURGE UNWANTED and UNUSED BLOCKS

You can remove a block reference from your drawing by erasing it; however, the block definition remains in the drawing's block definition table. To remove any **unused** blocks, dimension styles, text styles, layers and linetypes you may use the **Purge** command.

How to delete unwanted and unused blocks from the current drawing.

1. Select the **Purge** command using one of the following:

    Ribbon = None

    Application Menu = Drawing Utilities / Purge

    Keyboard = Purge <enter>

2. Select the + sign beside **Blocks**.

3. Select the block that you wish to purge.

4. Select the **Purge** button.

5. Select the **Close** button.

WHERE ARE BLOCKS SAVED?

When you create a Block it is saved **within the drawing you created it in**.

(If you open another drawing you will not find that block.)

MULTILEADER and BLOCKS

In Lesson 19 pages you learned about Multileaders and how easy and helpful they are to use. *I saved this command for this lesson because it works best with Blocks attached to the Leader.* In this lesson you will learn another user option within the Multileader Style Manager that allows you to <u>attach a **pre-designed Block**</u> to the landing end of the Leader. To accomplish this, you must <u>first create the style</u> and <u>then you may use it</u>.

Here are a few examples of multileaders with pre-designed blocks attached:

<u>STEP 1. CREATE A NEW MULTILEADER STYLE</u>

1. Select the **Multileader Style Manager** tool using

   Ribbon = Annotate tab / Leaders panel / ↘

2. Select the **New** button.

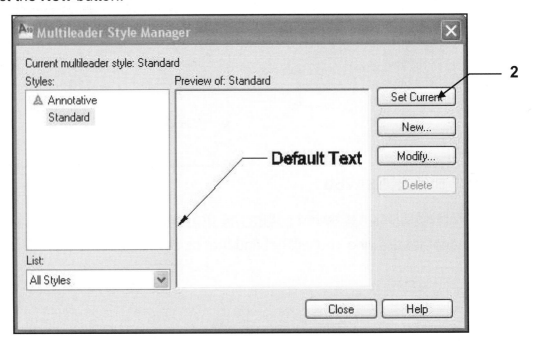

MULTILEADER and BLOCKS....continued

3. Enter **New Style Name**

4. Select a style to **Start with**:

5. Select **Annotative** box

6. Select **Continue** button.

7. Select the **Content** tab.

8. **Multileader Type**: Select **Block** from drop down menu.

9. **Source Block**: Select **Circle** from the drop down menu.
   Note: you have many choices here. These are AutoCAD pre-designed blocks with Attributes.

10. **Attachment**: Select **Center Extents**.
    Note: this selection works best with Circle but you will be given different choices depending on which Source block you select.

11. **Color**: Select **Bylayer**
    Bylayer works best. It means, it acquires the color of the current layer setting.

12. **Scale:** Select 1.000

13. Select **OK** button.

MULTILEADER and BLOCKS....continued

Your new multileader style should be displayed in the Styles list.

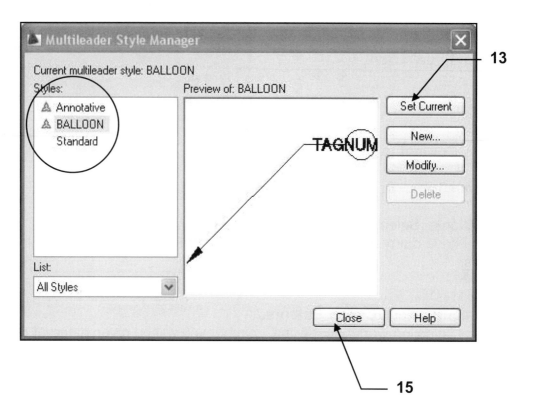

14. Select **Set Current**

15. Select **Close** button.

MULTILEADER and BLOCKS....continued

STEP 2. USING THE MULITLEADER WITH A BLOCK STYLE.

1. Select Layer **Symbol** from the Layer manager.

2. Select the **Style** from the Multileader drop down list

3. Select the **Multileader** tool.

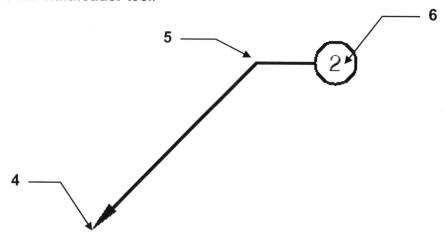

4. Specify leader arrowhead location or [leader Landing first/Content first/Options] <Options>: *place the location of the arrowhead*

5. Specify leader landing location: *place the location of the landing*

> *The next step is where the pre-assigned "Attributes" activate.*
> *Refer to the Advanced Workbook for Attributes.*

6. Enter attribute values
    Enter tag number <TAGNUMBER>: *type number or letter <enter>*

COLLECT MULTILEADER

In lesson 19 you learned how to ADD, REMOVE and ALIGN multileaders. Now you will learn how to use the **COLLECT** multileader tool.

If you have multiple Leaders pointing to the same location or object you may wish to COLLECT them into one Leader.

Example:

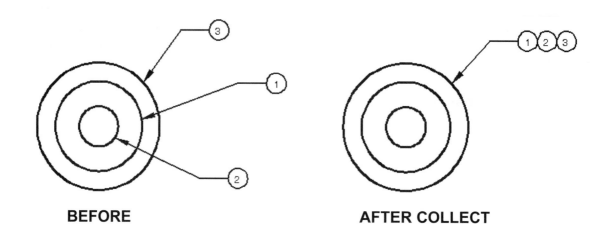

BEFORE   **AFTER COLLECT**

HOW TO USE COLLECT MULTILEADER

1. Select the Collect Mutileader tool

2. Select the Multileaders that you wish to combine.
   Note: Select them one at a time <u>in the order you wish them to display</u>.

   Such as (1, 2, 3, A, B, C, etc)

3. Place the combined Leader location. (**Ortho** should be **OFF**)

EXERCISE 29A

STEP 1

1. Start a **NEW** file using **My Decimal Setup.dwt.**

2. Select the **Model** tab

3. Draw the objects shown below approximately as shown. <u>Use Layer Object</u>.

4. Do not dimension

5. Create a block of each.

    A. **Important:** Don't forget to select a "**Base Point**" somewhere on each object.

    B. Select "**Delete**" so they disappear as they are made.

         ○ Retain
         ○ Convert to block
         ◉ Delete

    C. <u>Do not include the name</u> when selecting the objects for each block.

6. Save as **EX-29A**

EXERCISE 29A continued

STEP 2

1. Open **EX-29A** (If not already open)

2. Select the **A Size** tab

3. Draw the Network Diagram approximately as shown

4. Insert the Blocks you created in Step 1.

5. Add the Labels   use: Text Style: Class Style   Height: .125

6. Edit the Title Block

7. **Save** the drawing as: **EX-29A** again

8. **Plot** using page setup **Plot Setup A**

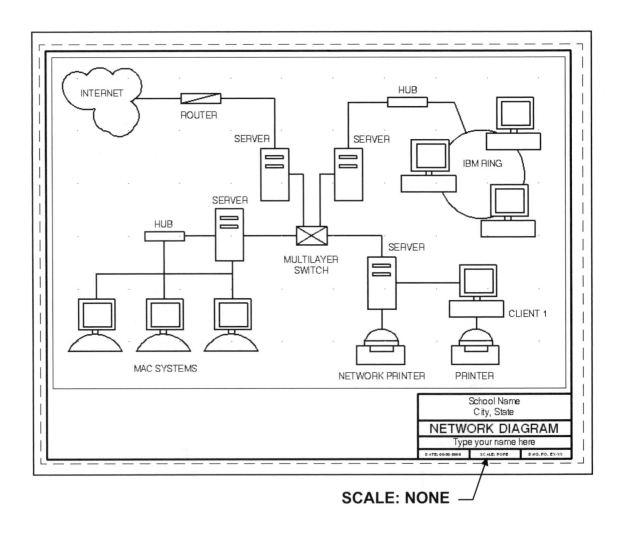

EXERCISE 29B

STEP 1

1. Start a **NEW** file using **My Feet-Inches Setup.dwt.**

2. Select the **Model** tab

3. Draw the objects shown below approximately as shown. **Use Layers specified**.

4. Do not dimension

5. Create a block for each.

   A. **Important:** Don't forget to select a "**Base Point**" on each object.

   B. Select "**Delete**" so they disappear as they are made.

   C. Use the Numbers for name. <u>Do not include the number</u> when selecting the objects for the block.

6. Save as **EX-29B**

   Note: Your blocks will appear much smaller and thinner than the blocks shown below. These have been enlarged for clarity.

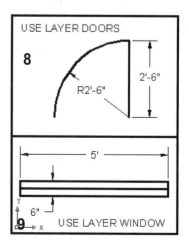

EXERCISE 29B continued

STEP 2

1. Open **EX-29B** (If not already open)

2. Select the **A Size** tab

3. Draw the Floor plan approximately as shown below inside the viewport.

   (Note: the viewport should already be scaled to 1/4" = 1')

   A. The walls are 6" wide.

   B. The space behind the door is 4" and counter depth is 24".

4. Insert the Blocks DOOR (8) on Layer Door

5. Insert the Block WINDOW (9) on Layer Window.

6. Dimension as shown using Dimension Style **Dim-Arch** (Your dimension text will appear different. I used text style Text-Classic because it is easier to read)

7. **Save** the drawing as: **EX-29B** again

EXERCISE 29B continued

STEP 3

1. Open **EX-29B** (If not already open)

2. Add the Electrical Blocks (1, 2, 3, 4 and 5) on Layer Electrical.

3. Add the Plumbing Blocks (6 and 7) on Layer Plumbing.

4. **Save** the drawing as: **EX-29B** again

Note: The dimension layer has been temporarily turned off so you can see the electrical and plumbing blocks easier.

EXERCISE 29B continued

STEP 4

1. Open **EX-29B** (If not already open)

2. Add the Wiring.

   A. Use Layer Wiring

   B. Use Arc, (Start, End, Direction)

3. Notice that your dashed lines do not appear like the example below.

   A. Type **LTS** (this is <u>Linetype Scale</u> setting)

   B. Type **.25**

4. **Save** the drawing as: **EX-29B** again

EXERCISE 29B continued

STEP 5

1. Open **EX-29B** (If not already open)

2. Erase the existing Viewport Frame. (The Floor plan will disappear)

3. Create **3 NEW** Viewports approximately as shown.

4. Adjust the scale of each Viewport.

5. Use **PAN** to find the area to view as shown below.

6. **Lock** each Viewport.

7. **Add** the dimensions in the 2 Viewports on the right
   by adding the current object scale. (Refer to page 28-3 and 28-4)

8. **Add** the Viewport Scale labels (1/8" ht) in Paperspace (Not Model space)

9. Make Viewport Frames plottable. (page 3-11)

10. **Save** as **EX-29B**   **Edit** Title Block   **Plot** using Plot Page Setup **Plot Setup A**

EXERCISE 29C

INSTRUCTIONS:
STEP 1

1. Start a **NEW** file using **My Decimal Setup.dwt.**

2. The Viewport should already be scaled to 1 = 1 and locked.

3. **Draw** the objects shown below.

4. Draw the **Break** line using Polyline and the Spline option (page 24-3)

5. Draw the Hatch using Pattern Ansi 31 and Layer Hatch.

6. Draw the Threads:

      A. Use Layer Threads

      B. Thread lines are .0625 distance apart.

7. Dimension as shown using Layer Dimension and Dimension Style = Dim-Decimal.

8. Save as **EX-29C**

EXERCISE 29C

STEP 2

1. **Freeze** the Dimension Layer.

2. **Assemble** the parts as shown below.

3. Use <u>Move, Mirror, Rotate, Trim, Erase</u> and any other command that you need.

4. The distance between the Jaws should be 1"

5. Save as **EX-29C**

| School Name |
| City, State |
| **STEP 2 (EX-30B)** |
| Type your name here |
| DATE:00-00-0000 | SCALE:1-1 | DWG. NO. EX 30 |

EXERCISE 29C

STEP 3

1. Draw the List of parts **in Paper space** above the title block as shown.

2. Use Multiline Text to enter text. **Justify** = Middle Center   **Text ht** = .125 and .062

3. Add Ballooned Leaders as shown **in Model Space**. Use Collect and Align.

4. Save as **EX-29C** and **plot** using plot page setup **Plot Setup A**

List of Parts (enlarged)

LEARNING OBJECTIVES

After completing this lesson, you will be able to:

1. Understand Parametric drawing
2. Apply Geometric and Dimensional constraints
3. Control the display of constraints
4. Plot Dimensional constraints

LESSON 30

Parametric Drawing

Parametric drawing is a method of assigning a constraint to an object that is controlled by another object. For example, you could put a constraint on line #1 to always be parallel to line #2. Or you could put a constraint on diameter #1 to always be the same size as diameter #2. If you make a change to #1 AutoCAD automatically makes the change to #2 depending on which constraint has been assigned.

There are two general types of constraints:

Geometric: Controls the relationship of objects with respect to each other. The Geometric constraints that will be discussed in this workbook are coincident, collinear, concentric, fix, parallel, perpendicular, horizontal, vertical, tangent, symmetrical and equal.

Example:

Dimensional: Dimensional constraints control the size and proportions of objects. They can constrain:
A. Distances between objects or between points on objects
B. Angles between objects or between points on objects
C. Sizes of arcs and circles.

Example:

Geometric Constraints

Overview of Geometric Constraints.

Geometric constraints control the relationship of objects with respect to each other. For example, you could assign the parallel constraint to 2 individual lines to assure that they would always remain parallel.

The following is a list of Geometric Constraints available and their icons:

You can apply geometric constraints to 2D geometric objects only. Objects cannot be constrained between model space and paper space.

When you apply a constraint, two things happen:
1. The object that you select adjusts automatically to conform to the specified constraint.
2. A gray constraint icon displays near the constrained object.

Example:

If you apply the **Parallel** constraint to lines A and B you can insure that line B will always be parallel to line A.

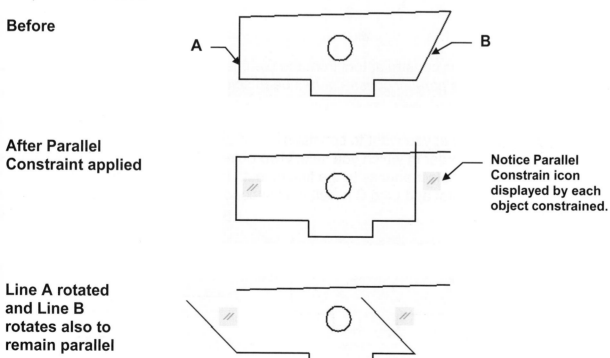

Before

After Parallel Constraint applied

Notice Parallel Constrain icon displayed by each object constrained.

Line A rotated and Line B rotates also to remain parallel

Geometric Constraints....continued

How to apply Geometric Constraints

The following is an example of how to apply the Geometric Constraint **Parallel** to 2D objects.
   The remaining Geometric Constraints are described on the following pages.

1. Draw the objects.
    Geometric constraints must be applied to <u>existing</u> geometric objects.

2. Select the **Parametric** tab.

3. Select the Geometric constraint tool from the Geometric panel.
    (In this example the parallel constraint has been selected.)

4. Select the objects that you want to constrain.
    In most cases the order in which you select two objects is important. Normally the second object you select adjusts to the first object. In the example shown below Line A is selected first and Line B is selected second. As a result Line B adjusts to Line A.

Notice Parallel Constrain icon displayed by each object constrained.

Geometric Constraints....continued

Coincident Constraint

A coincident constraint forces two points to coincide.

1. Draw the objects.

2. Select the Coincident tool from the Geometric panel.

3. Select the first Point.
   (Remember it is important to select the points in the correct order. The first point
   will be the base location for the second point)

First point

4. Select the second Point.
   (Remember it is important to select the points in the correct order. The second
   point will move to the first point)

Second point

The points are now locked together. If you move one object the other object will move
also and the points will remain together.

Both points coincide

Geometric Constraints....continued

Collinear Constraint

A collinear constraint forces two lines to follow the same infinite line.

1. Draw the objects.

2. Select the Collinear tool from the Geometric panel.

3. Select the first line.
   (The first line selected will be the base and the second line will move.)

4. Select the second line.
   (The second line will move in line with the first line selected)

The two lines are now locked in line. If you move one line the other line will move also and they will remain collinear.

Geometric Constraints....continued

Concentric Constraint

A concentric constraint forces selected circles, arcs, or ellipses to maintain the same center point.

1. Draw the objects.

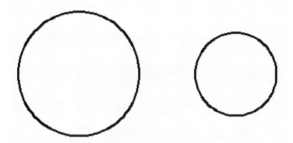

2. Select the Concentric tool from the Geometric panel.

3. Select the first circle.
   (The first circle selected will be the base and the second circle will move.)

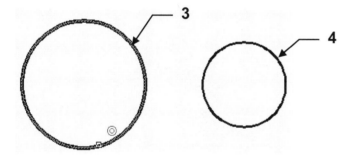

4. Select the second circle.

The second circle moves to have the same center point as the first object.

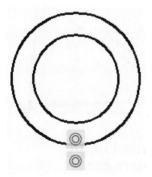

Geometric Constraints....continued

Fixed Constraint

A Fixed constraint fixes a point or curve to a specified location and orientation relative to the Origin (World Coordinate System, WCS).

1. Draw the object.

2. Select the Fixed tool from the Geometric panel.

3. Select the Fixed location.

The bottom left corner is now fixed to the specified location but the other three corners can move.

Fixed corner but the other corners can move using grips

Geometric Constraints....continued

Perpendicular Constraint

A Perpendicular constraint forces two lines or polyline segments to maintain a 90 degree angle to each other.

1. Draw the objects.

2. Select the Perpendicular tool from the Geometric panel.

3. Select the First object.
   (The first object will be the base angle and the second will rotate to become perpendicular to the first)

   3. First object

4. Select the Second object.

   4. Second object

The second object rotated to become perpendicular to the first object.

Geometric Constraints....continued

Horizontal Constraint
The Horizontal constraint forces a line to remain <u>parallel to the **X-axis**</u> of the current UCS

1. Draw the objects.

2. Select the Horizontal tool from the Geometric panel.

3. Select the <u>object</u> .

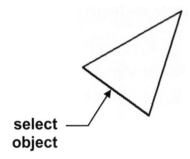

select object

The object selected become horizontal to the X-Axis of the current UCS.

Note: If the object is a Polyline or Polygon use the "2 point" option. The first point will be the pivot point.

2nd Point

1st point (Pivot point)

1st point (Pivot point)

2nd point

Geometric Constraints....continued

Vertical Constraint

The Vertical constraint forces a line to remain <u>parallel to the **Y-axis**</u> of the current UCS.

1. Draw the objects.

2. Select the Vertical tool from the Geometric panel.

3. Select the <u>object</u> .

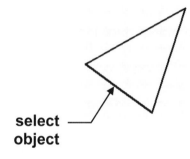

select object

The object selected become <u>vertical to the Y-Axis of the current UCS</u>.

Note: If the object is a Polyline or Polygon use the "2 point" option. The first point will define the pivot point.

nd Point

1st point
(Pivot point)

1st point
(Pivot point)

2nd point

Geometric Constraints....continued

Tangent Constraint
The Tangent constraint forces two objects to maintain <u>a point of tangency</u> to each other.

1. Draw the objects.

2. Select the Tangent tool from the Geometric panel.

3. Select the base object (Large Circle) and then the object (Line) to be tangent.

4. Select the base object (Large Circle) and then the object (Small Circle) to be tangent.

Geometric Constraints....continued

Symmetric Constraint

The Symmetric constraint forces two objects on a object to maintain <u>symmetry about a selected line.</u>

1. Draw the objects.

2. Select the Tangent tool from the Geometric panel.

3. Select the base line (1) then the Line (2) to be symmetrical and then the line (3) to be symmetrical about.

4. Select the base line (1) then the Line (2) to be symmetrical and then the line (3) to be symmetrical about..

Geometric Constraints....continued

Equal Constraint
The Equal constraint forces two objects to be equal in size. (Properties are not changed

1. Draw the objects.

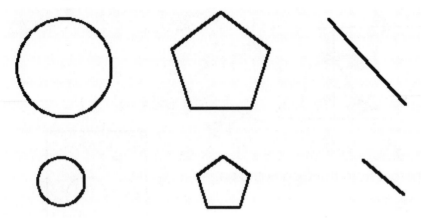

2. Select the Equals tool from the Geometric panel.

3. Select the base object (A) then select the object (B) to equal the selected base object

Geometric Constraints....continued

Controlling the display of Geometric constraint icons

You may temporarily Hide the Geometric constraints or you may show individually selected constraints using the Show and Hide tools.

Show All

This tool displays all geometric constraints.
Click on the tool and the constraints appear.

Hide all

This tool hides all geometric constraints.
Click on the tool and the constraints disappear.

Show

After you have selected the Hide All tool to make the constraints disappear, you may display individually selected geometric constraints.
1. Select the Show tool
2. Select the object
3. <enter>
The geometric constraints for only the selected objects will appear.

Dimensional Constraints

Overview of Dimensional Constraints
Dimensional constraints determine the distances or angles between objects, points on objects, or the size of objects.

Dimensional constraints include both a name and a value.

Dynamic constraints have the following characteristics:
    A. Remain the same size when zooming in or out
    B. Can easily be turned on or off
    C. Display using a fixed dimension style
    D. Provide limited grip capabilities
    E. Do not display on a plot

There are 7 types of dimensional constraints. (They are similar to dimensions)
    1. Linear
    2. Aligned
    3. Horizontal
    4. Vertical
    5. Angular
    6. Radial
    7. Diameter

The following is an example of a drawing with dimensional constraints. The following pages will show you how to add dimensional constraints and how to edit them.

Dimensional Constraints....continued

How to apply Dimensional Constraints

The following is an example of how to apply the dimensional constraint Linear to a 2D object.

1. Draw the objects.

2. Select the Parametric tab.

3. Select the Dimensional Constraint tool from the Dimensional panel.
   (In this example the Linear tool has been selected)

4. Apply the "Linear" dimensional constraint as you would place a linear dimension.
   a. Place the first point
   b. Place the second point
   c. Place the dimension line location

5. Enter the desired value or <enter> to accept the displayed value.
   (Notice the constraint is highlighted until you entered the value)

4C. dimension line location

5. Value highlighted temporarily

4A. First Point

4B. Second Point

6. The Dimensional constraint is then displayed with Name (d1) and Value (12.38).

Dimensional Constraints....continued

How to use Dimensional Constraints

The following is an example of how to use dimensional constraint to change the dimensions of a 2D object.

1. Draw the objects.
   (Size is not important. The size will be changed using the dimensional constraints.)

2. Apply Dimensional constraints
   Linear and Diameter.

Note: Dimensional constraints are very faint and will not print.

Continued on the next page...

Dimensional Constraints....continued

Now I will adjust the length and width using the dimensional constraints.

1. Double click on the d1 dimensional constraint.

2. Enter the new value for the length then <enter>

Note: The length increased automatically in the direction of the "2nd endpoint". If you want the length to change in the other direction you must apply the "geometric constraint Fixed" to the right hand corners. (See page 30-8)

3. Double click on the d2 dimensional constraint and enter the new value for the width then <enter>

Dimensional Constraints....continued

Parameter Manager

The Parameter Manager enables you to manage dimensional parameters. You can change the name, assign a numeric value or add a formula as its expression.

Select the Parameter Manager from the Parametric tab / Manage Panel

Name column: Lists all of the dimensional constraints. The order can be changed to ascending or descending by clicking on the up or down arrow.

Expression column: Displays the numeric value or formula for the dimension.

Value column: Displays the current numeric value.

Dimensional Constraints....continued

Parameter Manager Name cell

You may change the name of the dimension to something more meaningful. For example you might change the name to Length and width rather than d1 and d2.

1. Double click in the name cell that you want to change.

2. Type the new name and <enter>

Example:

The names in the Parameter Manager shown below have been changed. Notice the dimensional constraints in the drawing changed also.

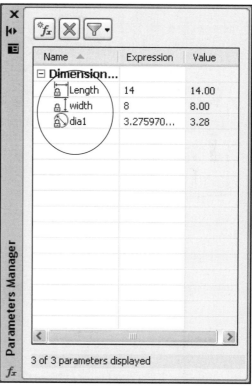

Dimensional Constraints....continued

Parameter Manager Expression cell

You may change the value of a dimensional constraint by clicking on the Expression cell and entering a new value or formula.

1. Open the Parameter Manager. (Refer to previous page)

2. Select the Expression cell to change (dia1 in this example)

3. Enter new value or formula.
   For this example: width / 2
   This means the diameter will always be half the value of d2 (width)

Now whenever the width is changed the diameter will adjust also.

Notice the Value cell displays the current value

Dimensional Constraints....continued

Adding User-defined parameters
You may create and manage parameters that you define.

1. Open the Parameter Manager

2. Select the "New user parameter" button.
   A "**User Variable**" will appear.

2

3. Enter a **Name** and an **Expression**.
   The Value cell updates to display the current value.

Note: With Imperial Units, the parameter manager interprets a minus or a
dash (-) as a unit separator rather than a subtraction operation. To specify
subtraction, include at least one space before or after the minus sign.
For example: to subtract 9" from 5', enter **5' -9"** rather than **5'-9"**

Dimensional Constraints....continued

Convert Dimensional constraint to an Annotational constraint

Geometric and Dimensional constraints do not plot. If you would like to plot them you must convert them to an Annotational constraint.

1. Select the constraint to convert. (Click on it once)

2. Right click and select **properties** from the list.

.
3. Select the Constraint Form down arrow and select Annotational

4. The Properties palette is populated with additional properties as the constraint is now an annotational constraint.

5. The annotational constraint will now plot.

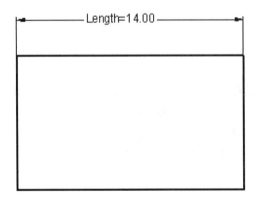

Dimensional Constraints....continued

Control the Display of Dimensional constraints
You may turn off the display of Dimensional constraints using the Show Dynamic Constraints button. Click once to turn OFF. Click again to turn ON.

Delete a Dimensional constraint.
To permanently delete a dimensional constraint select the Delete Constraints button and then select the constraint to delete.

EXERCISE 30A

INSTRUCTIONS:

1. Start a **NEW** file using **My Decimal Setup.**

2. **Draw** the Lines shown below on the left using random lengths and angles.

3. Create the shape on the right using the lines on the left and Geometric constraints.

4. Add the two Dimensional constraints.

5. Change the Dimensional constraints name as shown.

6. Create an Expression formula for the Width. (pg 30-22)

7. Add a <u>User defined parameter</u> **Area**

8. Change the Dimensional Constraints to Annotational. (pg. 30-24)

9. **Save** the drawing as: **EX30A**

EXERCISE 30B

INSTRUCTIONS:

Step 1

1. Open **EX-27G**

2. Select Layer **0**

3. Freeze layers: Dimension, Doors, Furniture and Windows

Step 2

1. Using the Polyedit / Join command (pg 24-3) join the lines that form the walls into 2 sets of polylines.

2. **Save** the drawing as: **EX30B**

Continue on to Step 3 on the next page

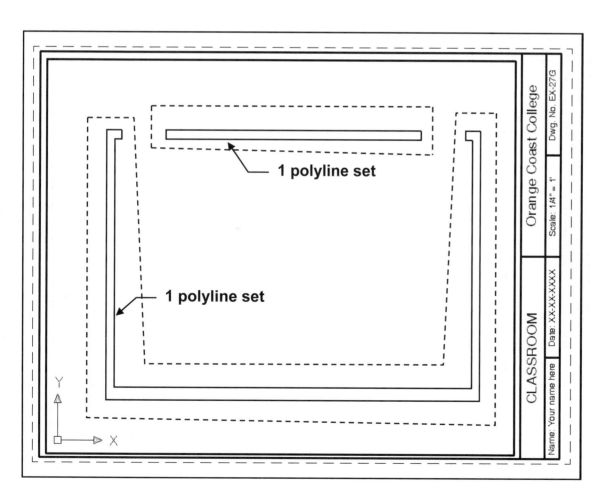

1 polyline set

1 polyline set

CLASSROOM

Orange Coast College

Name: Your name here | Date: XX-XX-XXXX | Scale: 1/4" = 1' | Dwg. No. EX-27G

EXERCISE 30B....continued

INSTRUCTIONS:

Step 3

1. Using Geometric constraints, make **all** inside lines **Parallel** to the outside lines.

   (Refer to page 30-3)

Note:

Your geometric icons may appear in different locations than the illustration below.

2. **Save** the drawing as: **EX30B**

Continue on to Step 4 on the next page

EXERCISE 30B....continued

INSTRUCTIONS:

Step 4.

1. Apply Dimensional constraints as shown below.

Note:

Notice where the 1st point is located. When placing the Length and Width dimensional constraints, it is important to select the 1st and 2nd points in the correct order. The objects will change size in the direction of the 2nd point.

2. **Save** the drawing as: **EX30B**

Continue on to Step 5 on the next page

EXERCISE 30B....continued

INSTRUCTIONS:

Step 5.

1. Change d1 to 31' (refer to page 30-19)

2. Change d2 to 22'

Note:

If you placed everything correctly d1 should have increased to the right and d2 down.
Do not be surprise if it does not work correctly the first time. If it did not, try to figure out
what you did wrong. You may not have made all of the lines parallel to each other or
maybe you did not place the 1st points of the dimensional constraints in the correct
location. If you master Geometric and Dimensional constraints it could be very helpful
in the future.

3. **Save** the drawing as: **EX30B**

APPENDIX A
Add a Printer / Plotter

The following are step-by-step instructions on how to configure AutoCAD for your printer or plotter. These instructions assume you are a single system user. If you are networked or need more detailed information, please refer to your AutoCAD Help Index.

Note: You can configure AutoCAD for multiple printers. Configuring a printer makes it possible for AutoCAD to display the printing parameters for that printer.

A. Type: **Plottermanager <enter>**

B. Select **"Add-a-Plotter"** Wizard

Add-A-Plotter Wizard
Shortcut
1 KB

C. Select the **"Next"** button.

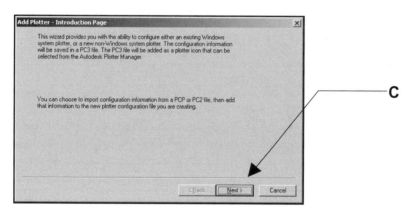

D. Select "**My Computer**" then **Next**.

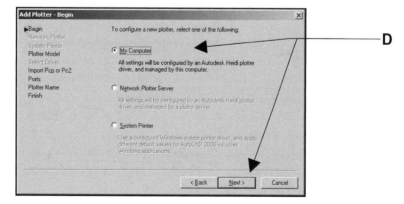

E. Select the **Manufacturer** and the specific **Model** desired then N**ext**.

   (If you have a disk with the specific driver information, put the disk in the disk drive
   and select "Have disk" button then follow instructions.)

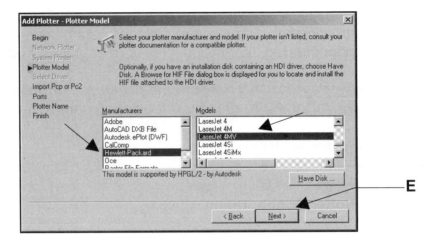

F. Select the **"Next"** box.

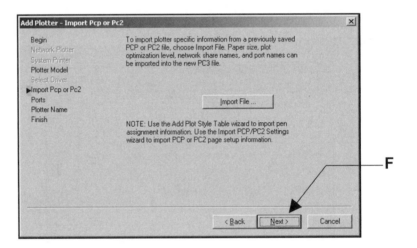

G1. Select **"Plot to a port"**. ——— G1
G2. Then select **"Next"**.

Select LPT1 o
if you are
configuring y
printer and it
attached to th
computer.
Not necessar
select any if
selected print
is not attache
to this
computer.

H. The Printer name that you previously selected should appear. Then select **"Next"**

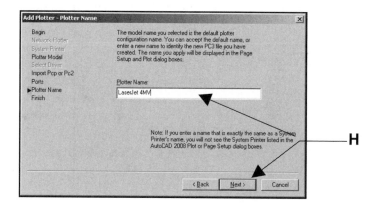

I. Select the **"Edit Plotter Configuration..."** box.

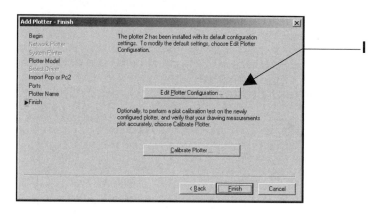

J. Select:
1. Device and Document Settings tab.
2. Media: Source and Size
3. Size: (Select the appropriate size for your printer / plotter)
4. OK box.

K. Select **"Finish"**.

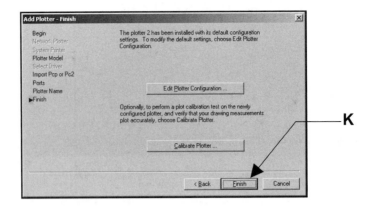

L. Type: **Plottermanager <enter>** again.

Is the printer / plotter there?

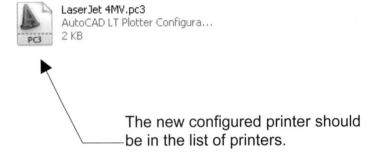

The new configured printer should be in the list of printers.

APPENDIX B
METRIC CONVERSION FACTORS

Multiply

Length	By	To Obtain
centimeter	0.03280840	foot
centimeter	0.3937008	inch
foot	0.3048a	meter (m)
foot	30.48a	centimeter (cm)
foot	304.8a	millimeter (mm)
inch	0.0254a	meter (m)
inch	2.54a	centimeter (cm)
inch	25.4a	millimeter (mm)
kilometer	0.6213712	mile [U.S. statute]
meter	39.37008	inch
meter	0.5468066	fathom
meter	3.280840	foot
meter	0.1988388	rod
meter	1.093613	yard
meter	0.0006213712	mile [U. S. statute]
microinch	0.0254a	micrometer [micron] (mm)
micrometer [micron]	39.37008	microinch
mile [U.S. statute]	1609.344a	meter (m)
mile [U. S. statute]	1.609344a	kilometer (km)
millimeter	0.003280840	foot
millimeter	0.03937008	inch

Multiply Length	By	To Obtain
rod	5.0292a	meter (m)
yard	0.9144a	meter (m)

Area

	By	To Obtain
acre	4046.856	meter2 (m2)
acre	0.4046856	hectare
centimeter2	0.1550003	inch2
centimeter2	0.001076391	foot2
foot2	0.09290304a	meter2 (m2)
foot2	929.0304a	centimeter2 (cm2)
foot2	92,903.04a	millimeter2 (mm2)
hectare	2.471054	acre
inch2	645.16a	millimeter2 (mm2)
inch2	6.4516a	centimeter2 (cm2)
inch2	0.00064516a	meter2 (m2)
meter2	1550.003	inch2
meter2	10.763910	foot2
meter2	1.195990	yard2
meter2	0.0002471054	acre
mile2	2.5900	kilometer2
millimeter2	0.00001076391	foot2
millimeter2	0.001550003	inch2
yard2	0.8361274	meter2 (m2)
fathom	1.8288	meter (m)

Symbols of SI units, multiples and sub-multiples are given in parentheses in the right-hand column.

Appendix B-2

APPENDIX C - DYNAMIC INPUT

To help you keep your focus in the "drawing area", AutoCAD has provided a command interface called **Dynamic Input**. You may input information within the Dynamic Input instead of on the command line.

The **First point** displays the (Absolute: X, Y) distance from the Origin.
Enter the **X** dimension, <u>press the **Tab** key</u>, enter the **Y** dimension then **<enter>**.

The **Second and Next points** display (Relative: Distance and Angle) from the last point entered. Enter the **distance**, <u>press the **tab key**</u>, enter the **angle** then **<enter>**.

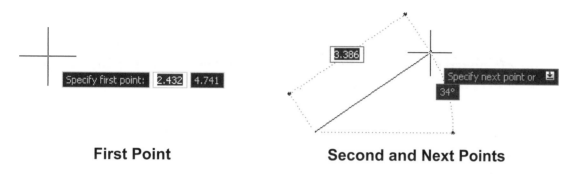

| **First Point** | **Second and Next Points** |

Note: *Some users find Dynamic Input useful some find it distracting.* *After completing this lesson you decide if you want to use it.* *It is your choice.*

How to turn Dynamic Input ON or OFF

Select the **DYN** button on the status bar or use the F12 key.

DYNAMIC INPUT has 3 components

1. **Pointer Input**
2. **Dimensional Input**
3. **Dynamic Prompts**

You may control what is displayed by each component and turn each ON or OFF.

Refer to the following pages for more detailed descriptions.

POINTER INPUT

Pointer Input is only displayed for the **first** point.
When Pointer Input is enabled (ON) and a command has been selected, the location of the crosshairs is displayed as coordinates in a tooltip near the cursor.

Example:
1. Select the **LINE** command.

2. Move the cursor.

The 2 boxes display the cursor location as X and Y coordinates **from the Origin**. (Absolute coordinates)

3. You may move the cursor until these coordinates display the desired location and press the left mouse button or enter the desired coordinate values in the tooltip boxes instead of on the command line.

How to change POINTER INPUT settings

1. Right-click the **Dyn** button on the status bar and select **Settings**.

2. In the Drafting Settings dialog box select the **Dynamic input** tab.

3. Under Enable Pointer Input select the **Settings** button.

4. In the Pointer Input Settings dialog box select:

    Format: (select one as the display format for second or next points)

    > **Polar or Cartesian format**

    > **Relative or Absolute coordinate**

    Visibility: (select one for tooltip display)

    - o   **As Soon As I Type Coordinate Data**. When pointer input is turned on, displays tooltips only when you start to enter coordinate data.

    - o   **When a Command Asks for a Point**. When pointer input is turned on, displays tooltips whenever a command prompts for a point.

    - o   **Always—Even When Not in a Command**. Always displays tooltips when pointer input is turned on.

5. Select **OK** to close each dialog box.

DIMENSIONAL INPUT

When Dimensional Input is enabled (ON) the tooltips display the distance and angle values for the **second** and **next points**.

The default display is <u>Relative Polar</u> coordinates.

Dimensional Input is available for: **Arc, Circle, Ellipse, Line** and **Polyline**.

Example:
1. Select the <u>Line</u> Command.

2. Enter the 1st point.

3. Move the cursor in the direction you desire and enter the **distance, Tab key,** and the **angle <enter>.**

Angle

41°

2.658

Specify next point or

Distance

How to change DIMENSIONAL INPUT settings

1. Right-click the **Dyn** button on the status bar and select **Settings**.

2. In the Drafting Settings dialog box select the **Dynamic input** tab.

3. Under <u>Dimension Input</u>, select the **Settings** button.

4. In the Dimension Input Settings dialog box select:

    Visibility: (select one of the following options)

    o   **Show Only 1 Dimension Input Field at a Time**. Displays only the distance dimensional input tooltip when you are using grip editing to stretch an object.

    o   **Show 2 Dimension Input Fields at a Time**. Displays the distance and angle dimensional input tooltips when you are using grip editing to stretch an object.

    o   **Show the Following Dimension Input Fields Simultaneously**. Displays the selected dimensional input tooltips when you are using grip editing to stretch an object. Select one or more of the check boxes.

5. Select **OK** to close each dialog box.

DYNAMIC PROMPTS

When Dynamic Prompts are enabled (ON) prompts are displayed.

You may enter a response in the tooltip box instead of on the command line.

Example:
1. Select the <u>Circle</u> command.

2. Press the right mouse button for command options or you may also press the down arrow to view command options.
    Select an option by clicking on it or press the up arrow for option menu to disappear.

How to change the COLOR, SIZE, or TRANSPARENCY of tooltips

1. Right-click the **Dyn** button on the status bar then select **Settings**.

2. In the Drafting Settings dialog box select the **Dynamic input** tab.

3. At the bottom of the dialog box select **Drafting Tooltip Appearance** button.

4. Select "**Colors**" button. Under <u>Context</u>: select <u>2D Model Space</u>. Under <u>Interface Elements</u>: select <u>Drafting Tool Tip</u> or <u>Drafting Tool Tip Background</u>. Select a color from the Color drop down list. Then select <u>Apply and Close</u> button.

    (I like a light gray)

5. Under **Size**, move the slider to the right to make tooltips larger or to the left to make them smaller. The default value, 0, is in the middle.

    (I like 2, a little larger than the default)

6. Under **Transparency**, move the slider. The higher the setting, the more transparent the tooltip.

    (I like 25, a little more transparent than the default)

7. Under **Apply To**, choose an option:

    o   **Override OS Settings for All Drafting Tooltips**. Applies the settings to all tooltips, overriding the settings in the operating system.

    o   **Use Settings Only for Dynamic Input Tooltips**. Applies the settings only to the drafting tooltips used in Dynamic Input.

8. Select **OK** to close each dialog box.

INDEX

D

AutoCAD 2010 trial software

The enclosed disk contains a fully functioning 30-day trial version of AutoCAD 2010 software. This software has been provided to give you a preview of how easily AutoCAD 2010 installs and is used in conjunction with this Exercise Workbook. After 30 days, from the day you install the software, you will not be able to open the program. At that time you will have the opportunity to activate the software by purchasing the actual software license.

IMPORTANT: The 30-day trial version of AutoCAD can be loaded only once on a single computer. This applies to all 30-day trials regardless of where you obtained them, including from other Exercise Workbooks by Cheryl Shrock.

If you are attending a school, check with your instructor for educational discounts.
Or visit: www.autodesk.com for additional purchasing information.

Make efficiency a daily part of the job with AutoCAD® 2010 software. Meticulously refined with the drafter in mind, AutoCAD 2010 propels day-to-day drafting forward with features that increase speed and accuracy while saving time. Annotation scaling and layer properties per viewport minimize workarounds, while text and table enhancements and multiple leaders help deliver an unmatched level of aesthetic precision and professionalism.

AutoCAD 2010 system requirements
Refer to page Intro-9

How to install AutoCAD:

Note: AutoCAD 2010 can coexist with earlier versions of AutoCAD and AutoCAD LT.

1. Insert the AutoCAD 2010 Disk into your computer's drive.

2. In the AutoCAD Media Browser, on the install tab, click Stand-Alone Installation.

3. Under Install AutoCAD 2010, click Install.

4. On the Autodesk Setup page, click OK to install required support components.

5. In the AutoCAD Installation wizard, follow the directions on each page.

6. When the installation is complete you will be prompted to:
   "Authorize AutoCAD" or "run AutoCAD without authorizing."

   Select "**run AutoCAD without Authorizing**"

Note: This trial software will stop working 30 days from the date of installation.